Handbook of Saline and Alkali Soils :
Diagnosis, Reclamation and Management

The Authors

Dr. S K Gupta, currently, INAE Visiting Professor at ICAR-NDRI, Karnal, obtained B. Tech (Agril. Engg.) and Master of Agricultural Engineering from the Punjab Agricultural University, Ludhiana and Ph.D. from JNTU, Hyderabad, Telangana. Dr. Gupta is Fellow of the National Academy of Agricultural Sciences, Indian National Academy of Engineering and Indian Society of Agricultural Engineering. He has served as Head, Indo-Dutch Network Project, Head, Division of Irrigation and Drainage Engineering and Project Coordinator (AICRP on Management of Salt Affected Soils and Use of Saline Water in Agriculture). After superannuation, Dr. Gupta worked as Emeritus Scientist (ICAR) and INAE Distinguished Professor. Dr. Gupta has been engaged in conducting researches in surface and subsurface drainage for land reclamation, irrigation water management, hydrology of salt affected soils, and use of saline water in agriculture. He has published more than 150 research papers and 15 books. He has been bestowed with Rafi Ahmad Kidwai Award by ICAR. Besides he is recipient of several awards from NAAS, ISAE, IE (India), CBIP and MoWR among other organizations.

Dr Parbodh Chander Sharma, presently working as Director, Central Soil Salinity Research Institute, Karnal (India), was earlier program leader for improvement of rice, wheat, Indian mustard and chickpea for salt tolerance and high yield as Head, Division of Crop Improvement. He has made significant contributions in understanding the physiological mechanisms under salinity stress in different crops and was involved in developing salt tolerant high yielding genotypes of Indian mustard (CS 56, CS 58 and CS 60), rice (CSR 46, CSR 56 and CSR 60) and wheat (KRL 283), besides improving three popular mega varieties of rice (Pusa 44, PR 114 and Sarjoo 52) for salinity tolerance by introgression of *SALTOL* QTL following molecular marker assisted backcross breeding. He has made significant contribution in developing crop and resource management practices for sustainable future cereal-based systems. He has been awarded fellowships by three professional societies. He is the President of the Indian Society of Soil Salinity and Water Quality since 2016. He has more than 100 research publications in peer reviewed research journals.

Dr. Suresh Kumar Chaudhari is presently working as Assistant Director General (Soil and Water Management) in Indian Council of Agricultural Research, New Delhi. He obtained his graduation and post-graduation from Jawaharlal Nehru Agricultural University, Jabalpur and Ph.D. from Mahatma Phule Agricultural University, Rahuri. He had specialized trainings in the area of management of saline and sodic soils from United States Soil Salinity Laboratory, USA and University of Rostock, Germany. He has developed deep insights and fundamental and conceptual models of soil-water movement through salt affected soils. He has worked on various scientific positions in different Institutes of ICAR. He is a Fellow of prestigious professional academies and societies like National Academy of Agricultural Sciences, Indian Society of Soil Science etc. Dr. Chaudhari is a recipient of ICAR-Rafi Ahmed Kidwai Award, Outstanding Achievement Award in 2010, 12th International Congress Commemoration Award and TWAS Medal in 2009.

Handbook of Saline and Alkali Soils : Diagnosis, Reclamation and Management

S.K. Gupta
P.C. Sharma
S.K. Chaudhari

SCIENTIFIC PUBLISHERS

Published by:

Scientific Publishers (India)
5-A, New Pali Road, P.O. Box 91,
Jodhpur – 342 001 (India)
E-mail: info@scientificpub.com
Website: www.scientificpub.com

ISBN: 978-9
978-9

ISBN:9789388812276
9 789388 812276
Price: 2650/-

Printed in India

PREFACE

Salt affected soils have been a part and parcel of the landscape since time immemorial. Although people knew that excessive salts in soils reduce productivity and the land value, only half hearted efforts were made to reclaim these soils. It is attributed to the scanty information available to the people in those times and investments on reclamation being uneconomical and risky. Groups of researchers were working at various institutions around the globe to unravel the fundamental principles of land reclamation and develop technologies to reclaim these soils. Voluminous information was generated in countries like India, USA, USSR, Australia, Hungary and many other nations. Establishment of United States Regional Salinity Laboratory in 1937, which was rechristened as US Salinity Laboratory in 1951 provided a real impetus to the research activities on salt affected soils. The Laboratory in 1954 brought out the publication titled 'Diagnosis and Improvement of Saline and Alkali Soils' which is popularly known as USDA Handbook 60. It had a global impact on the reclamation and management of salt affected soils. Even today, it is a treasured possession of people dealing with salt affected soils in any manner.

India, during all these years, was also charting its own course through groups of people engaged in understanding the formation, reclamation and management of salt affected soils. The scattered information on the subject available with these groups was compiled by Agarwal and Gupta. It was published by the Indian Council of Agricultural Research (ICAR) in 1968 as Technical Bulletin 15. At about the same time, ICAR-Central Soil Salinity Research Institute (CSSRI) was established on 1st March 1969 with its headquarters at Karnal, Haryana, close to the village Munak, where first report of soil salinity emerged in 1855. Establishment of ICAR-CSSRI proved to be a landmark in the history of salinity research in India.

ICAR-CSSRI, within a year of its establishment, successfully tested the alkali land reclamation technology. Over the years, the technology was passed on to the state governments as a result of which about 2.07 M ha of alkali lands have been reclaimed, till now. Besides increased production of food grains, it has huge socio-economic impact. Simultaneously, the institute also made significant dent in the reclamation of saline soils, development of salt tolerant crop varieties, reclamation of coastal saline soils, reclamation and management of Vertisols, use of saline and alkali water and reuse of domestic wastewater. The institute has researched almost all facets of salt affected soils and poor quality water and widely applied reclamation and management technologies under real field situations.

The information included in this handbook has emerged from the concerted research efforts of the entire staff of the CSSRI, and its regional Research Stations. Equally good contributions have come from several active groups working at various

State Agricultural Universities of India. We also borrowed few salient research results from the international literature available at our command. In spite of our deep desire to acknowledge each and everyone contributing to these efforts, we refrained from naming individuals mainly for the two reasons. One the list is long and two we do not wish to miss anyone which in spite of our best efforts is likely to happen. As a testimony to the efforts made by the scientists, we have included the most relevant literature with appropriate citation. We decided to place on record our indebtedness to the esteemed directors of CSSRI for their administrative acumen and guidance. We take this opportunity to place on record our hearty thanks to Drs. D.R. Bhumbla, Late J.S.P. Yadav, I.P. Abrol, N. T. Singh, N.K. Tyagi, Gurbachan Singh, D.K. Sharma and current director P.C. Sharma. We also acknowledge with thanks the efforts put in by various reviewers namely Drs. M.J. Kaledhonkar, D.S. Bundela, A.K. Rai, Bhaskar Narjary, B. Nirmalendu and Ashim Datta. They unhesitatingly accepted our request and critically reviewed the chapters in a time bound manner.

The CSSRI is in its 50th glorious year. Several national and international recognitions have been bestowed upon the institute during these years. Although a matter of great satisfaction, the institute has a long way to tread. Many complex and challenging problems are awaiting solutions. We are confident that the new generation of scientists equipped with new knowledge and capabilities and able guidance of senior people will be able to resolve these issues. We hope this handbook will help initiate the entry of the new generation of scientists into the fascinating world of salinity research. We also hope and believe that the handbook will serve as a reference book to the scientists engaged in salinity research and to the officers of the line departments engaged in land reclamation.

S.K. Gupta
P.C. Sharma
S.K. Chaudhari

CONTENTS

Chapter - 1

GENESIS AND DISTRIBUTION OF SALT AFFECTED SOILS

Salt affected soils have been a part of the global landscape. While the history of salts in soils is as old as the formation of the soils, its impact on crop productivity has also been known to the mankind since the beginning of agriculture. References to the '*Urvar*' (productive) and the '*Anurvar'* (unproductive) soils in the old Hindu Literature is a testimony to this statement. '*Anurvar'* soils were also referred as '*Usar*' soils, a Sanskrit word derived from '*Ushtra*'. Lands having excess salts such as neutral NaCl or sodic ($NaHCO_3$, Na_2CO_3), were called *'Reh'* by the geologists in mid-nineteenth century to characterize the appearance of salt efflorescence at the surface of lands. The word 'sodic' is of Arabic origin meaning ash-like. Arabs named these soils as such since the surface soil looked like calcined wood ash. The salt affected soils included in this Handbook owe their special characters to excess soluble salts and/or high exchangeable sodium percentage or both on the soil surface and in the root zone. While excessive soluble salts are harmful to plants because plants cannot draw water from the salty soil solution, these salts can also increase the ESP of the soil if the soluble constituents consist largely of sodium salts. The latter is a more permanent manifestation of excess salts since the problem persists even if soluble salts are removed from the soil. Crop growth is adversely affected under both the situations unless remedial interventions are undertaken to reclaim these soils.

TERMINOLOGY OF ALKALI SOILS

In a compilation of works from India and University of California, Hilgard (1886, 1906) lamented that the term "alkali soils" was being applied almost indiscriminately in USA and India to all soils containing an unusual amount of soluble mineral salt as most were having pH exceeding 7.0. Hilgard (1906) therefore, divided salt affected soils into two classes namely 'white alkali' and 'black alkali'. While both these soils have excess of sodium, the former has neutral salts such as NaCl and Na_2SO_4 while the later has Na_2CO_3 and $NaHCO_3$ type of salts. In the former, soil pH and ESP are not high even though SAR may be quite high. It is expected that SAR will reduce during leaching with or without slight increase in pH. These soils are also termed as high SAR saline soils or simply the saline soils. The black alkali soils are the real sodic soils, the name being derived from the black spots often seen on the surface of these soils, being the evaporates of salts and organic matter that imparts black colouration. The nomenclature however, has become obsolete although many people continue to use these terms.

Sigmond (1927) categorized the soils as: saline soils, salty alkali soils, leached alkali soils, degraded alkali soils and regraded alkali soils. Overstreet et al. (1951) visualized alkali soils of saline, sodic, and saline-sodic types. Sodic soils, having high amount of exchangeable sodium, were characterized as “sodic” alkali soils. Authors justified this nomenclature stating that they have preserved the historical meaning of the term "alkali" as a general class name denoting soils with high sodium contents either in the form of free salts or in the adsorbed state. At the same time, the little-used adjective "sodic" (containing sodium) has been chosen to denote undesirably high amounts of adsorbed sodium. Richard (1954) of the Soil Salinity Laboratory of United States classified these soils, into non-saline alkali soils, saline-alkali soils and saline soils. The preamble of the Handbook says that in deference to past usage, the term “alkali soil” is employed to refer to soils that have a high exchangeable-sodium-percentage; and “saline soil” is used in connection with soils having a high value for the electrical conductivity of the saturation extract. The term sodic soil is not even included in the glossary of terms given in the Handbook. Nomenclature Committee appointed by the Board of Collaborators of the US Salinity Laboratory [1958] recommended the use of the term ´sodic´ instead of ´alkali´ as previously proposed by the Salinity Laboratory. Use of this terminology has since been approved by the Soil Science Society of America [1960].

Clearly, terms sodic and alkali soils have been and are being used interchangeably in the literature. On the other hand, it has also emerged that sodic soils are not always alkali as these can be acidic/saline as well (Rengasamy, 2006, 2010). Degraded sodic soils reported in deltaic regions of West Bengal (India) also have ESP more than 15 but pH in the range of 5.0-7.0 (Chakravarti and Chakravarti, 1957). In several cases reported in India and elsewhere saline soils have a high sodicity (a high SAR), but they are not sodic soils and have usually a good infiltration capacity. Rengasamy (2006) elaborately categorized various salt affected soils as acidic–saline soils, neutral saline soils, saline soils, alkaline–saline soils, acidic saline–sodic soils , acidic–sodic soils, non-alkali sodic–acid soils, neutral saline–sodic soils, neutral sodic soils, non-alkaline sodic and strongly sodic neutral soils, alkali saline–sodic soils and alkali–sodic soils.

In the Indian context, the most commonly used USDA classification of salt affected soils has been adopted according to which salt affected soils are classified in 3 groups’ namely saline, alkali and saline-alkali soils. Since there is no unanimity on the nomenclature of alkali soils, both the terms are being used on the wishes of the writer(s). Even in the same write-up, one can find both the terms. CSSRI prefers to call these soils alkali until some unanimity is reached through internal discussions involving experts of various disciplines and organizations. As such, this Handbook will use the term alkali except at places where literature is referred and the writer has used the term sodic for alkali soils.

Besides, what is stated in the foregoing paragraphs, several other kinds of salt affected soils having similar issues have been reported. Some such soils are: degraded sodic soil, a non-saline-alkali soil in some areas of western United States and India having ESP above 15 but pH of the surface soil as low as 6, also designated as acid-sodic soils, calcium chloride rich soils, soils containing excessive exchangeable magnesium (magnesium solonetz), gypsiferous soils containing large amount of gypsum or soils with the presence of a structural 'B' horizon. The magnitude of these soils is limited and in some cases the problems may not be as severe as in the two broad categories of saline and alkali soils. Therefore, the discussions in this Handbook are restricted to the saline and alkali soils.

NATURE AND CLASSIFICATION

The most commonly used USDA classification of salt affected soils, classifies salt affected soils in 3 groups' namely saline, alkali and saline-alkali soils (Richards, 1954).

Table 1.1 Classification of salt affected soils on the basis of chemical characteristics

Class	EC_e (dS m^{-1})	ESP[§]	pH_2[$]
Saline soils	> 4*	< 15	< 8.5
Saline-alkali soils	> 4	>15	Variable
Alkali soils	< 4	>15	>8.5

*Present literature sometimes gives this value as 2.0 dS m^{-1} mainly because many sensitive crops such as gram and vegetables suffer at this level of salinity. [$]If pH_s is measured, the value is taken as 8.2 (Abrol et al., 1980)

Saline Soils

The soils having predominantly neutral salts such as chlorides and sulphates of sodium, calcium, and magnesium having electrical conductivity (EC) > 4 dS m^{-1}, pH_2 less than 8.5 but not less than 7.0, with the exception of some coastal saline soils like those found in Kerala, India, and ESP < 15 are categorized as saline soil (Table 1.1). In some gypsiferous soils, like those found in Gujarat and Rajasthan states of India, calcium sulphate is the dominant salt. In most saline soils, the whole profile is salt laden, although the salt content in the profile gradually decreases with depth. Strongly saline soils also show a white 1 to 5 cm thick salt layer on the soil surface during the dry period. It is why these soils have been referred as white alkali, the term that has become almost obsolete.

Within saline soils, following three distinct groups have been identified.

- Inland saline soils having shallow water table, water quality being saline in nature. These soils suffer from water logging and salinity (Fig. 1.1). These soils are mostly encountered in irrigation commands of India.

- Saline soils formed as a result of irrigation with saline groundwater. These soils normally have deep water table. However, in certain situations the water table may be shallow. These soils generally occur in arid and semi-arid regions.
- Coastal saline soils where the problem is due to ingress of sea water resulting in high salinity and temporary flooding. The climate is generally humid to semi-arid/arid.

Due to excessive neutral salts, these soils are usually flocculated and have good physical properties. The structure is however, unstable due to lack of organic matter. As the soils are mostly uncultivated, a hard layer develops below the fluffy salty surface layer due to the compaction caused by the hydrostatic pressure of the flood water during rainy season, the prolonged fallowing, and the trampling by animals. The conditions like periodic submergence or water logging and the presence of shallow brackish groundwater are the severe limitations, as these cause further salinization. Slightly saline soils are moderately cultivated with some salt resistant crops like barley, wheat, sugarcane, oats, *berseem* and *sesbania.* Plants growing on such soils may appear stunted with thickened leaves of dark green colour. On saline lands, salt tolerant grasses appear during the monsoon season but large barren spots are common. Saline soils are potentially productive due to abundant nutrient reserves and ideal moisture retention and transmission characteristics following their reclamation. In local parlance of different states these soils are termed as *Thur, Uippu, Lona, Shora, Soula, Pokhali , Khar and Kari.*

Fig. 1.1 A view of water logging and soil salinity in western Jamuna canal command of Haryana

Alkali Soils

Alkali soils contain excess of salts like bicarbonates, carbonates and silicate of sodium, capable of alkaline hydrolysis and have sufficient exchangeable sodium to interfere with growth of most crops (Fig. 1.2). Appreciable amounts of carbonate ions can be present only at pH values of 9.5 or higher. Following reaction between Na_2CO_3 and H_2O explains the high pH in soils containing carbonates.

$$2\ Na^+ + CO_3^{-2} + 2H^+ = 2\ Na^+ + 2\ OH^- + H_2CO_3 \qquad (1.1)$$

Carbonic acid H_2CO_3 is unstable and produces H_2O and CO_2 (CO_2 escaping into the atmosphere). This explains the remaining alkalinity (or rather basicity) in the form of soluble sodium hydroxide and the high pH or low pOH.

The pH_2 of alkali soils is > 8.5, EC_e is variable and ESP >15 (Table 1.1). There is a white encrustation on the surface during dry months, which often shows dark/black colour on wetting. It is because dispersed organic matter accumulates at the surface of poorly drained areas. As water evaporates, it imparts a black colour to the surface on wetting, hence the name "black alkali" soils. This term like white alkali has also become almost obsolete. Dispersion of the clay and organic fractions occurs because of high ESP. When wet, the soil become sticky and on drying it is cloddy, hard, compact and difficult to work with. Rainwater does not move down and remains standing for a long period. The soil presents a barren and desolate look. The excessive sodium carbonate in the soil solution causes the precipitation of calcium of the exchange complex as insoluble calcium carbonate, thus causing a high amount of exchangeable sodium. The ESP values of 80-90 or more are common in the surface layers of these soils. Generally, ESP decreases with depth. These soils have excessive soluble salts in the upper 0-30 cm soil surface layer. The impact of alkalinity is revealed through surface crusting, reduced seedling emergence, increased run-off and erosion risk, sparse ground cover and low organic matter adversely impacting soil aeration and microbial activity. These soils in various states of India are commonly referred as *Usar, Rakkar, Bara, Chopan, Kari.*

Fig. 1.2 A view of a barren alkali land

Saline- Alkali Soils

These soils have a combination of harmful quantities of salts and a high content of exchangeable sodium, which interferes with the growth of crop plants. The pH_2 of these soils is > 8.5, EC_e is > 4 dS m^{-1} and ESP >15 (Table 1.1). As long as excess salts are present, these soils have all the features of a saline soil. The soils remain in flocculated condition and may have pH around 8.5. If reclamation procedures are used that do not include application of amendment, they become alkali and show higher pH upon leaching. These soils in various states of India are commonly referred as *Usar, Kallar, Karl, Chopan, Bari, Reh, Choudu and Kshar.*

DIAGNOSTIC CRITERIA

On the basis of experiences in India, the visual and physico-chemical characters that distinguish these soils are given in Table 1.2 and Table 1.3.

Table 1.2 The visual observations to distinguishing between a saline and an alkali soil

Observation	Saline soil	Alkali soil
Rain water stagnation	Relatively lesser period	Stagnation for longer period
Stagnated water type	Clear	Dark coloured, muddy, soapy
Surface salt crust	White salt efflorescence	Dark brown or ash coloured clay crust
Groundwater	Saline, sometimes with high Sodium adsorption ratio (SAR)	Usually good, occasionally with high residual sodium carbonate (RSC)
Presence of nodules	Gypsum may be present at some depth	$CaCO_3$ nodules are present
Natural vegetation (grasses)	*Cressa cretica, Cyperus rotundus, Chloris pallida, Sporobolus pallidus, Haloxylon salicornicum, Acluropus lagopoides, Zygophyllum simplex, Dichanthium annulatum, Sueda furuticosa, Butea monosperma*	*Sporobolus marginatus, Desmostachya bipinata, Suaeda maritma, Diplachne fusca, Dichanthium annulatum, Cynodon dactylon, Chloris barbata, Brachiaria mutica, Kochia indica, Panicum antidotale*

Table 1.3 Characteristics of saline and alkali soils

Characteristics	Saline soil	Alkali soil
EC_e (dS m^{-1})	4.0 or more	Variable
ESP	<15	>15
pH_s	< 8.2	8.2 or more
Chemistry of soil solution	Dominated by chlorides and sulphates of sodium, calcium and magnesium.	Measurable to appreciable quantities of salts capable of alkaline hydrolysis, e.g. Na_2CO_3 and $NaHCO_3$ always present
Physical properties	In the presence of excess neutral soluble salts, clay fraction is flocculated	High exchangeable sodium and pH result in the dispersion of clay and the soils have an unstable structure
Permeability	Permeability of soils to water and air and other physical characteristics are generally comparable to normal soils	Permeability of soils to water and air is restricted. Physical properties of the soils become worse with increasing levels of exchangeable sodium/pH
Main adverse effect on plants	High osmotic pressure of soil solution resulting in reduced uptake of water	Alkalinity of soil solution, chiefly through the dispersive effect of excess exchangeable sodium resulting in poor physical properties. Interactive effect of pH on nutrition and calcium deficiency
First step for reclamation	Removal of excess electro-lytes through leaching coupled with drainage, if waterlogged	Lowering or neutralization of high pH using amendments

SAR as a Diagnostic Criterion

Initially SAR of the soil solution was not used as a diagnostic criterion. Later on a correlation between SAR and ESP for many soils was established and it emerged that SAR 13 approximately equalled ESP 15 of these soils. Since it is easier to measure SAR than ESP, SAR to categorize salt affected soils gained momentum especially in Australia. It emerged that soils having SAR < 13 should be categorized as saline while those having SAR >13 as alkali or saline-alkali depending upon the EC_e of the soil. In the Indian context, this classification criterion failed to classify the soils as both the saline and alkali soils can have high SAR. There are cases where saline soils may have much higher SAR than an alkali soil. These soils in India have been categorized as high SAR saline soils. Since the criterion has come to stay in the International literature, additional calculations were suggested to categorize soils as saline or alkali especially from management point of view (Table 1.4, Chhabra, 2004). Soils that have the ratio of either $(2CO_3 + HCO_3) / (Cl + 2SO_4) > 1$ or $Na/ (Cl + 2SO_4) > 1$ in mol.m^{-3}, should be treated as alkali and require chemical amendments for reclamation. When soils have both these ratios < 1, then, irrespective of their pH and SAR, these should be treated as saline. Under such situations, both SAR and EC_e decrease simultaneously on leaching.

Table 1.4 Classification of high SAR soils as saline or alkali

Soil category	Guidelines
Saline	In the saturated paste extracts, the ratio of soluble ions $(2CO_3 + HCO_3)/(Cl + 2SO_4) < 1$ and $Na/(Cl + 2SO_4)$ should be <1 when expressed in mol m^{-3}
Alkali/sodic	In the saturated paste extracts, the ratio of soluble ions $(2CO_3 + HCO_3)/(Cl + 2SO_4)$ should be >1 or $Na/(Cl + 2SO_4)$ should be >1 when expressed in mol m^{-3}

FORMATION OF SALINE AND ALKALI SOILS

A number of cations sodium (Na^+), calcium (Ca^{2+}) and magnesium (Mg^{2+}) and to a smaller degree potassium (K^+), anions chloride (Cl^-), sulphate (SO_4^{2-}) and bicarbonate (HCO_3^-) and to a smaller degree carbonate (CO_3^{2-}) are found in soils in various proportions. Amongst other ions, soil solution may contain ammonium (NH_4^+) and nitrate (NO_3^-). Amongst dominant anions are chloride (Cl^-) and sulphate (SO_4^{2-}) while dominating cations are Na^+, Ca^{2+} and Mg^{2+}. The quantum and relative proportions of these anions and cations have great impact on the chemical characteristics of these soils. Besides, relative proportion also governs the way these soils are reclaimed or managed.

Soil Salinization

Soluble salts that accumulate in the soils are derived from various sources and follow different pathways. The main processes of salt accumulation are:

Weathering of rocks and lack of natural leaching

All soils contain some salts as these are derived from primary minerals, which form the original and the most important source of salts in the soils. During the chemical weathering processes comprising of hydrolysis, hydration, solution, oxidation and carbonation, various constituents like Ca^{2+}, Mg^{2+} and Na^+ are gradually released and made soluble. Since geologic materials are highly variable in their elemental composition, some materials having higher salts than others, causes the variations in salt content as well as kinds of salts encountered in the soils. This kind of salinity is often referred as primary salinity. It is exuberated in arid and semi-arid regions where low rainfall and high evaporation prevails. The low precipitation in these regions is insufficient to leach down the soluble weathered products resulting in salt accumulation in the soil. Even if the salts move with rainwater, the restricted downward movement does not allow these to move deeper in the profile. In dry weather, when high temperatures prevail, the salts again move up with the water and are brought up to the surface where they are deposited as the water evaporates. On the contrary in humid regions, salts are leached to greater depths, which move on to the sea. As a result, saline soils are practically non-existent in humid regions, except in river deltas and other low-lying lands near the sea where soils are subjected to sea water inundation.

Unscientific use of poor quality water

Similar to soils, water also contain salts the nature and total salt concentration dictated by the kinds of geologic formations through which the runoff has passed. When salty waters are used for irrigation salts accumulate over a period of time especially when natural drainage is either inadequate or provision of sufficient drainage is not made before initiating irrigation (See Chapter 3).

Shallow water table and secondary soil salinization

As the irrigation proceeds without adequate drainage, the groundwater begins to rise because of groundwater accretion through leakage, seepage and deep percolation losses often encountered in surface irrigation schemes. In most irrigation projects in India, it has happened especially during initial years when the water is in plenty and farmers have the tendency to use it in excess. Once the water table reaches within 1-2 m of the soil surface, water and salts move upward as a result of capillary action. Besides the salts in the water, salts present in the soil profile also move upward into the root zone or even to the soil surface. At this stage water is either transpired or evaporated while salts remain in the root zone or at the soil surface resulting in soil salinization. This process of soil salinization has a great economic bearing as the soils in the irrigation commands turning saline impacts the productivity of irrigated lands. At times, soils go out of cultivation to render huge investments made on irrigation development as unproductive.

Salt transport from uplands

Alluvial material brought down by rivers also carries weathered rocks and the salts in it. It may be especially so if the parent material consists of marine deposits that were laid down during earlier geologic periods and have since been uplifted. Deposits of silt and salt in the plains may be one of the several causes of soil salinization.

Ingress of sea water

The ocean may be the source of salts in coastal regions as saline sea water enters the land near the coast during high tides. Sea water inundation may also be caused from other processes in operation in coastal regions. Besides, use of sea water for irrigation along the creeks results in soil salinization.

Salt laden sand blown by sea winds

In arid regions near the sea, appreciable amount of salts are blown by wind year after year, which are deposited on the soil surface. Salts so deposited are not easily washed back to the sea or leached to the lower soil horizon because of limited rainfall. These soils in due course of time are salinized to adversely impact the productivity. This process of soil salinization is in operation in the Kutch region of Gujarat.

Anthropogenic activities

Anthropogenic activities such as addition of inorganic fertilizers or land disposal of many waste products add considerable amount of salts to the soils. Although one may not add enough salts through such activities that may cause soil salinization yet coupled with other processes such activities positively enhance the risk of soil salinization.

Genesis of Alkali Soils: Global Experiences and Viewpoints

The term sodication was first used by the Soil Science Society of America for pedogenic processes that increase the exchangeable Na^+ content of the soil. Like soil salinity, soil alkalinity is also categorized as primary and secondary or anthropogenic. Primary alkalinity results from natural processes in which Na^+-rich sedimentary materials or rocks release Na^+ upon weathering. Water, rich in carbon dioxide of atmospheric or biological origin, also acts as active chemical weathering agent that releases appreciable quantities of the bicarbonates, the ion responsible for soil alkalization. Secondary alkalinity is a consequence of changes in land use or other anthropogenic interventions, which increase sodium saturation of the exchange complex or add bicarbonates and carbonates to the soils. In the presence of excessive sodium in the soil solution whether due to primary or secondary sources, calcium and magnesium ions are replaced with sodium ions on the exchange complex resulting in high soil ESP.

Most explanations for site specific formation of alkali/sodic soils are based on the principles enunciated by Kovda *et al.* (1973) and Hardy and Eugster (1970). Kovda stated that mixed salts while migrating from one place to another fractionate the constituents in a manner that the least soluble salt are retained close to the source, whereas the most soluble and mobile salts move to accumulate at the farthest end. Hardy and Eugster suggested that in an evaporating water containing mixed salts, first to precipitate is calcite followed by gypsum or sepiolite (magnesium silicate) depending upon the ionic environment. The residual solution is made up of more soluble salts. These two theories together suggest that the predominant salts near the site of weathering in the presence of carbon dioxide will be carbonates and bicarbonates of calcium, magnesium, potassium and sodium albeit with low concentrations. As the water moves from the more humid to the arid regions, the salts concentration increases. During evaporation, sodium salts separate from calcium salts because of their differential solubility. The soluble sodium salts concentrate high in the soil profile. As the soil solution is further concentrated with sodium, the percentage of sodium on the exchange complex continue to increase resulting in dispersion of clay and organic matter.

Most theories also contend that soil salinization is the primary process before soil sodication occurs. Compositions of the soluble salts that accumulate at the soil salinization stage determine whether soil sodication will occur or not. Sodication of the exchange complex is bound to take place in situations where sodium predominates. Desalinisation of sodicated soil is usually accompanied by the formation of Na_2CO_3 and pronounced deflocculation.

Whitting and Janitzky (1963) postulated that sodium carbonate may also be formed as a result of reduction of sulphate ions under anaerobic conditions in the presence of organic matter. The following reaction takes place.

$$Na_2SO_4 \rightarrow Na_2S \rightarrow Na_2CO_3 + H_2S \quad (1.2)$$

The sulphide gas leaves the system.

Sodic soils of Egypt in Wadi-Tumilat, Ferhash and Wadi-El-Natroun are reported to have formed by desalinization in the absence of enough divalent cations in some parts of the Nile Delta, by high carbonate and bicarbonate water in Wadi-Tumilat and by denitrification and sulphate reduction under anaerobic conditions in Wadi-El-Natroun (El-Gabaly, 1971). Irrigation using groundwater containing carbonate and bicarbonate is one of the chief contributing factors in the formation of sodic soils in many regions (Qureshi and Barret-Lennard, 1998). In Pakistan, sodicity in 2.3 M ha area has been attributed to this reason. Beek and van Breemen (1973) pointed out that highly sodic soils can develop in a closed basin with an excess of evaporation over precipitation if the inflowing water has a positive residual sodicity. Sodication may also occur when saline groundwater containing residual sodicity moves up by capillary action to the soil surface leaving behind the salts.

Genesis of Alkali Soils: Indian Experiences and Viewpoint

Several theories on the formation of alkali soils have been advanced in India. Most of these theories are based on location specific formation of alkali soils. Nonetheless each one of these throws enough light on various processes that may be in operation either singly or in combination during the soil alkalization/ sodication.

Bhargava *et al.* (1980) while tracing the origin and accumulation of salts in Indo-Gangetic plains of India discounted some of the theories discussed in the foregoing section. They attributed alkaline hydrolysis of feldspar in the uplands plains of Haryana for the origin of sodium. Weathering of alumino-silicate minerals through carbonation was attributed to yield solutions of carbonates and bicarbonates in addition to silica and alumina.

$$\text{Na-primary mineral} + H_2O \rightarrow \text{H-silicate clay} + NaOH \quad (1.3)$$

$$\text{H-silicate clay} + H_2O \rightarrow Al_2O_3.H_2O + H_2SiO_3 \quad (1.4)$$

$$NaOH + CO_2\ (\text{air}) \rightarrow NaHCO_3 \quad (1.5)$$

$$NaHCO_3 \rightarrow Na_2CO_3 + CO_2 + H_2O \quad (1.6)$$

These reaction products migrated with the subterranean and surface waters. The relative differences in relief between parts of the plains and Siwaliks and within the plains facilitated the accumulation of salt laden run-off to accumulate in the depressions. Relief and climate together facilitated seasonal flooding and evaporation of the accumulated run-off resulting in formation of sodic (alkali) soils in the form of localized stretches within the micro-depressions. It appears that sodic soils of Haryana, Punjab, Uttar Pradesh, Bihar and Madhya Pradesh have been formed through the direct process of sodification commencing from the soil surface. It is why, these soils differ from the solonetz (sodic soils) occurring in the parts of Europe, former USSR and USA. These soils are thought to have evolved through the sequential processes of salinization, sodification and desalinization.

Presence of sodic soils in hilly areas having high rainfall, udic moisture regime and steep slopes in several parts of Punjab, Haryana and Himachal Pradesh cannot be explained through the process described in the previous paragraph (Kumar, 1992). Quoting the earlier research on weathering of soil minerals, he observed that feldspars including albites become highly stable in high pH soil environment and therefore, feldspar weathering is not possible at such high pH (Kittrick, 1971, Lindsay, 1979). Further, the alumino-silicate weathering as the source of salts cannot explain the presence of appreciable amount of chlorides and sulphates that always occur in these soils with sodium carbonates. Therefore, Kumar (1992) suggested that the salts might have come from the Himalayas and specifically from the marine environment of Tethyan sea during Miocene to Eocene periods (25-40 million yr). After many geological upheavals these materials might have been transported by ancient river system in late Pleistocent to Recent period (1 million yr) and deposited to form the present day alluvial plains (Kumar *et al.,* 1993, 1994). Recent finding of salt, silt, sodium and smectite impregnated alluvium underneath normal soils in south-west Punjab, which is similar to the sodic parent material occurring in Siwalik Hills, confirms the transported nature of salinity.

Dhir (1998) argued that perfect fractionation as postulated by Kovda *et al.* (1973) rarely occurs in nature. He postulated that a variety of sources such as rain water, salts of aeolian nature and release of salts from silicate weathering might have resulted in the formation of salt affected soils in semi-arid tracts of India. He visualized that sodium chloride in the soils and groundwater of the Indo-Gangetic plains might have come from the Himalayan catchments (from both geological sources and rainfall), sulphate from rock formations in the hills and alkaline earth carbonate mainly from the weathering of plagioclase and scarce ferromagnesian minerals in rocks, soils and sediments. The sodium carbonate, a dominant constituent of sodic soil is mainly from the weathering of minerals in soils and sediments of the plains. He agreed with the contention postulated in the previously mentioned theories that distant sources have played a major role in the formation of saline and alkali soils.

Datta *et al.* (2002) observed that trona alone or possibly a mixture of trona and thermonatrite represents the end product of evaporative concentration of seasonal flood waters mixed with the local groundwater drawn up within the soils by capillary action. In the process of evaporation, calcite precipitation occurs below the surface, whereas trona and thermonatrite form on the surface. Crystallization of salts raises the pH of near-surface soil to 10.5. This observation is based on the fact that efflorescent crust that develops on less vegetated portions of the Indo-Gangetic plains of Uttar Pradesh (India) consists predominantly of the sodium carbonate minerals trona ($Na_2CO_3.NaHCO_3.2H_2O$) with minor thermonatrite ($Na_2CO_3.H_2O$). The authors concluded that flooding and evaporation cycles is a possible cause of the formation of this mineral and of alkali soils in the flood plains (Datta *et al.*, 2002).This theory is also supported by the theory proposed by DeSigmond (1925). According to this theory, sodic soils may originate under arid and semi-arid conditions, where in the past there has been a temporary abundance of humidity in the soil intersected by dry periods.

Both pedogenic and non-pedogenic calcium carbonate (PC and NPC) are present in Vertisols irrespective of the ecosystems to which they belong. While NPCs are part of the parent material of Vertisols, dissolution of NPCs and recrystallization of dissolved Ca^{2+} ions results in the formation of PC. The dissolution of the NPC to release Ca^{2+} ions and their recrystallization as $CaCO_3$ (PC) is the prime chemical reaction of the pedogenesis of sodic Vertisols. It is also confirmed through the observation of Pal *et al.*, (2000) that Vertisols of arid and semi-arid regions contain more PC in their soil control section (SCS) than those of sub-humid climates. Formation of PC is the prime chemical reaction responsible for the increase in pH, the decrease in the Ca/Mg ratio of exchange site with depth and in the development of sub-soil sodicity (Srivastava *et al.*, 2002).

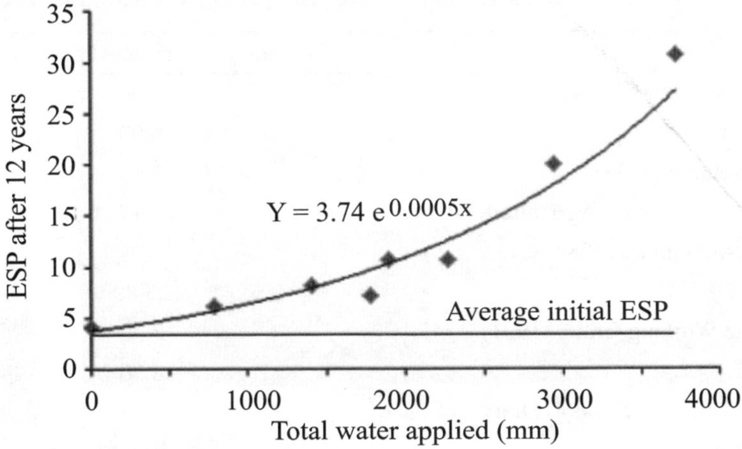

Fig. 1.3 Effect of alkali water application of soil alkalization
(Data source: Choudhary, 2017)

Most studies in India support the view that irrigation with alkali water having high RSC can be one of the causes of formation of alkali soils. The increasing ESP of the soil following increasing application of high RSC water (EC 1.4 dS m^{-1}, SAR 14.9 and RSC 10.1 meq L^{-1}) to cotton-wheat over a period of 12 years is depicted in Fig. 1.3. This process of soil sodification is operative in many parts of Haryana, Uttar Pradesh and several other states.

EXTENT AND DISTRIBUTION OF SALT AFFECTED SOILS

Using the FAO/UNESCO soil map of the world (1970-1980), Szabolcs (1979) and later FAO (http://www.fao.org/soils-portal/soil-management/management-of-some-problem-soils/salt-affected-soils/more-information-on-salt-affected-soils/en/) estimated that around 831 M ha area is salt affected out of which 397 M ha is saline and 434 M ha is sodic. In India salt affected soils occur extensively in different agro-ecological and soil zones of the country, particularly the arid, semi-arid and the sub-humid coastal regions. In the past, different organizations and authors have arrived at different estimates of salt affected soils in India due to the method of obtaining data, characterization guidelines used and several other factors (Table 1.5). The extent varied from a low value of 2.05 to 23.8 M ha. According to recent estimates, the salt affected soils occupy about 6.73 M ha in India (Table 1.6, Mondal *et al.*, 2011). The States most affected are Uttar Pradesh, Gujarat, Rajasthan, West Bengal and Andhra Pradesh (Fig. 1.4). It is feared that several million hectares of productive land in the newly irrigated areas will soon turn into wet deserts unless adequate preventive steps are taken. According to CSSRI (CSSRI, 1997, CSSRI, 2007, CSSRI, 2011, CSSRI, 2015) salt affected soils are projected to occupy 16.2 M ha area in 2050 (Fig. 1.5).

Table 1.5 The extent of salt affected soils as estimated by various workers

Source	Extent of salt affected area (M ha)
ICAR Seminar, 1959 (Shimla)*	5.00
Uppal (1966)	6.00
Abrol and Bhumbla (1971)	7.00
National Commission on Agriculture (1976)	7.17
Ministry of Agriculture (1984-85)	9.08
Abrol *et al.* (1988)	23.80
Report of the Working Group (1991)	3.30
Singh (1992)	8.56
Ministry of Agriculture (1994) Draft	6.32
Suman (1995)	10.60

*As reported by Agarwal and Gupta (1968)

Table 1.6 State-wise extent and distribution of salt affected soils in India

Sl. No.	State	Saline soils (ha)	Sodic soils (ha)	Total (ha)	Total (lakh ha)
1	Andhra Pradesh	77598	196609	274207	2.74
2	Andaman & Nicobar	77000	0	77000	0.77
3	Bihar	47301	105852	153153	1.53
4	Gujarat	1680570	541430	2222000	22.22
5	Haryana	49157	183399	232556	2.32
6	Karnataka	1893	148136	150029	1.50
7	Kerala	20000	0	20000	0.20
8	Maharashtra	184089	422670	606759	6.07
9	Madhya Pradesh	0	139720	139720	1.40
10	Odisha	147138	0	147138	1.47
11	Punjab	0	151717	151717	1.52
12	Rajasthan	195571	179371	374942	3.75
13	Tamil Nadu	13231	354784	368015	3.68
14	Uttar Pradesh	21989	1346971	1368960	13.69
15	West Bengal	441272	0	441272	4.41
	Total	2956809	3770659	6727468	67.27
	Total (M ha)				**6.73**

Source: Mondal *et al.*, 2011

ESTABLISHMENT OF CSSRI

In 1967, Indian Council of Agriculture Research appointed a Consultant Team to suggest the plan for research on water use and soil management towards meeting India's food shortages. The team after a thorough study of the problems recommended that to meet immediate food production goals, researches on applied or adaptive nature should begin immediately. ICAR should coordinate on water use and soil management research and training activities. In states, Agricultural Universities should carry out researches that have regional significance. To provide essential technological information for the water and soil management component of the 'Package of Practice', several research projects were recommended. A specific recommendation was made to undertake salinity research in the country. It recommended establishment of a coordinated drainage-salinity pilot research projects at one national center and others at regional centers (Black soils-Maharashtra or Madhya Pradesh; mixed Red and Black soils- Karnataka or

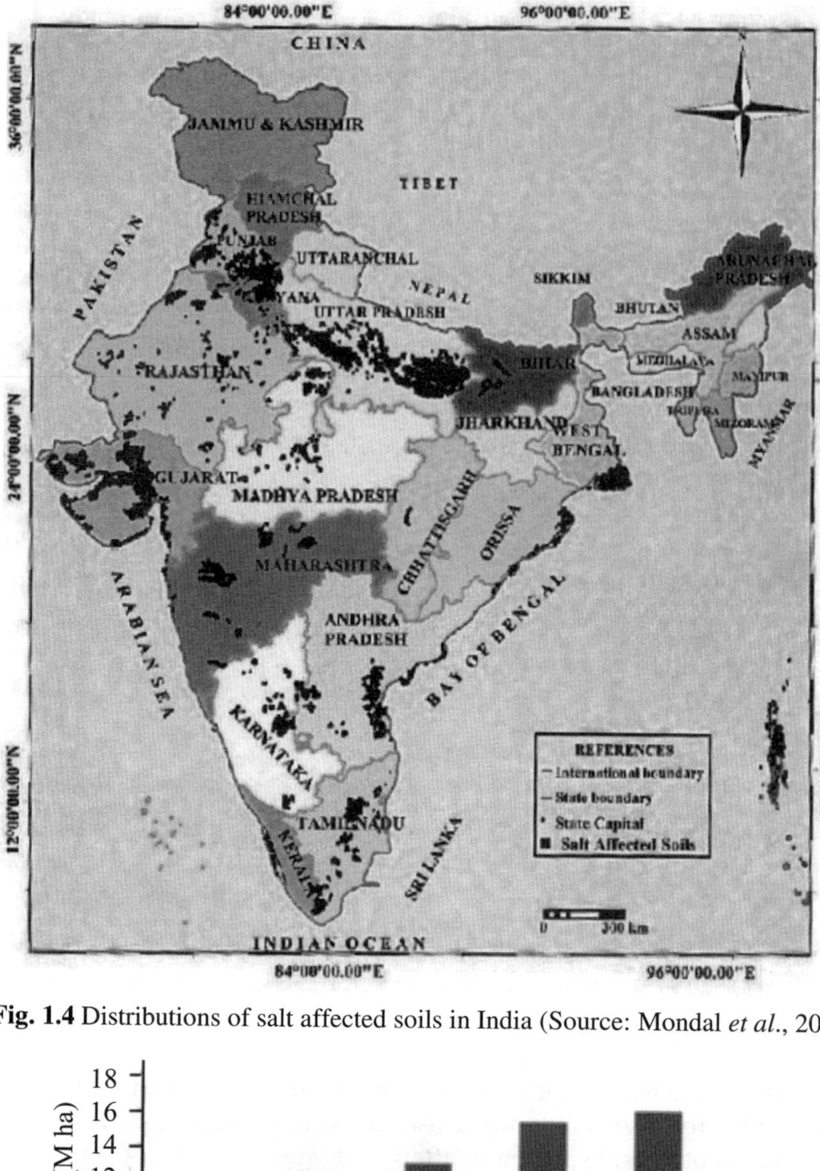

Fig. 1.4 Distributions of salt affected soils in India (Source: Mondal *et al.*, 2011)

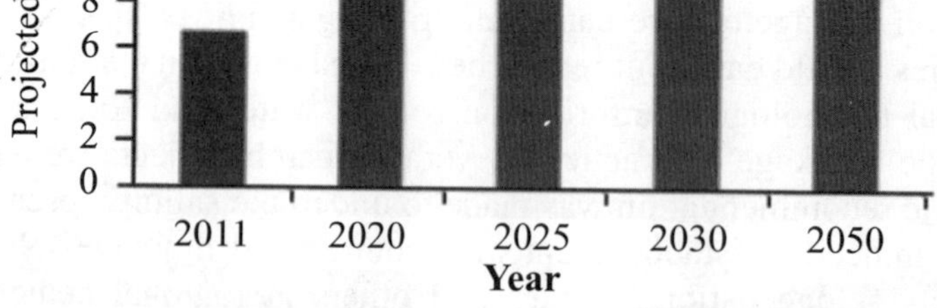

Fig. 1.5 CSSRI projections of the extent of salt affected soils in India

Odisha and Coastal alluvial soils- Odisha or West Bengal). These were mandated to evaluate the application of principles of drainage and management of salinity and sodicity in cultivated soils of India. In states other than those selected for the centers and where salt problems are acute, appointment of an Indian Scientist having in depth knowledge to aid cultivators was suggested. Establishment of a Central Water Management Research Center at the Indian Agricultural Research Institute to study water management problems of national significance also formed a part of these recommendations.

As a consequence of these recommendations, ICAR launched in 1968 an All India Coordinated Scheme for 'Studies on Soil Salinity and Water Management' at different locations and established Central Soil Salinity Research Institute (CSSRI) at Karnal in 1969. It is just by coincident that CSSRI happens to be located on the banks of the Western Jamuna Canal where first reports of salinity appeared during the middle of 19th century. The village Munak is just 30 km away from the campus of the institute. Another Coordinated scheme on 'Use of Saline Water in Agriculture' came in operation in 1972 at 5 centers in the country, which is now extended to 12 centers including 4 volunteer centers. With the establishment of this vast network, solutions to these menacing problems of water logging and soil salinity were expected to pave the path for sustainable irrigated farming in future. This handbook briefly describes basic and applied researches conducted in India over the last 50 years suitably connected to the past work done on this aspect. The authors believe that the outcomes as revealed through various technological and socio-economic parameters so far are in tune with the expectation from the institute.

■■■■

Chapter - 2

NATURE OF SALINE AND ALKALI SOILS

Physical, chemical and biological properties of the soils undergo a sea change upon salinization and/or alkalization. The knowledge of these properties serves as a basis for diagnosis, reclamation, and management of saline and alkali soils. As such, systematic investigations are undertaken to determine these properties, being a pre-requisite to design a reclamation programme. This chapter looks at important chemical, physical and biological properties of saline and alkali soils. Some of the chemical properties included in this chapter are listed in Table 2.1 while the physical and biological properties that form the subject matter of discussion are listed in Table 2.2.

Table 2.1 Soil chemical properties of salt affected soils

Analysis	Units	What it measures?	Interpretation of test data
Electrical conductivity (EC)	dS m^{-1}	Ability of soil solution to conduct electric current	Higher the soil or water EC, more the dissolved ions it contains An indicator of potential salinity of the soil/water
pH	Measured on a scale of 0–14	Acidity or alkalinity	An indicator of potential soil reaction Assesses whether pH is favorable for target crop Indicates potential solubility of soil minerals and nutrients
Exchangeable cations	meq $100g^{-1}$	Sodium (Na) Calcium (Ca) Magnesium (Mg) Potassium (K)	Concentration of cations adsorbed on negatively charged clay and silt particles (inorganic) and soil organic constituents Used to estimate exchangeable sodium percentage (ESP) Helps to assess gypsum requirement of alkali soils
Exchangeable sodium percentage (ESP)	Percentage of the total cation exchange sites occupied by sodium	Degree of sodicity or alkalinity	Used to determine gypsum requirement for treating alkali soils

Analysis	Units	What it measures?	Interpretation of test data
Cation exchange capacity (CEC)	c mol(p+) kg^{-1} or meq $100g^{-1}$soil	Capacity of soil to hold positively charged ions (cations)	For the same ESP/SAR, more gypsum needed to adjust SAR and ESP of soils having higher CEC
Calcium carbonate equivalent	%	Percentage of soil by weight that is un-dissolved limestone	If calcium carbonate is present, it may not be economical to adjust pH below 8. Presence of $CaCO_3$ is necessary for the success of acidifying soil amendments such as elemental sulphur or sulphuric acid
SAR	$(mmol\ L^{-1})^{1/2}$	Sodium (Na) Calcium (Ca) Magnesium (Mg)	Diagnosis of saline/alkali soils or water, gypsum may be required in high SAR soils to maintain infiltration rate, a indicator of ESP in some cases
Nutrients	mg L^{-1}, mg kg^{-1} or ppm	Macro, secondary and micro nutri-ents present in the soil	Determine the productive pote-ntial of soils, need to apply and to decide nutrient application rates

Table 2.2. Physical and biological parameters to assess the physical/biological status of the soils

Physical	Biological
Soil depth	Organic matter
Bulk density and porosity	Soil microbial biomass carbon
Infiltration rate	Nitrogen mineralization
Hydraulic conductivity	Soil enzymes assay
Soil water holding capacity	Bio-diversity
Dispersion coefficient	Crop yield
Clay swelling	
Water stable aggregates and structure stability	
Soil crusting	
Soil cracking	

SOIL SAMPLING

Soil salinity is among the most variable properties of soils, salt concentration varying greatly with horizontal or vertical distance as well as with time. Evidently, analysis of a single soil sample drawn from one place has little value for the appraisal of soil salinity/alkalinity. For a general appraisal of soil salinity in a given area, the average salt content of a number of samples will only provide a reliable information, larger the number of samples, the more accurate the appraisal. On the other hand, drawing and analysis of too many samples may increase costs, the cost of chemical analysis being greater than the cost of taking a sample. Clearly, a well-designed sampling plan is required to obtain reliable data for appraisal at a reasonable cost. Primary samples are drawn to prepare composite samples followed by reduction in the size of the sample using 'Quartering Technique' (See chapter 7).

For most chemical analysis disturbed representative samples are used. On the other hand physical properties are determined either in the field or on undisturbed and disturbed samples depending upon the kind of measurement and the accuracy required (See chapter 8).

CHEMICAL PROPERTIES

Electrical Conductivity (EC) and Soluble Salts

The total salt content of soil solution or irrigation water is expressed in milligrams per liter (mg L^{-1}) and millimole per liter (mmol L^{-1}) or electrolyte conductivity expressed in dS m^{-1} at 25°C. Earlier, it used to be expressed as TDS (total dissolved salts), which is the sum of the anions and cations present in units of ppm. Electrolyte conductivity is by far the simplest representation and is a measure of the ease with which electric current will flow through the soil water or the irrigation water. EC is reciprocal of the electrical resistivity, the resistance in ohms of a conductor, metallic or electrolytic, which is 1 cm long and has a cross sectional area of 1 cm^2. Hence, electrical conductivity was formerly expressed in reciprocal ohms per cm or mhos per cm and now in SI units it is expressed in dS m^{-1} (Table 2.3). It is equivalent to the earlier unit expressed in millimhos cm^{-1} (mmhos cm^{-1}) (1dS m^{-1} = 1 mmhos cm^{-1}).

Table 2.3 Conversion factors for soil/water salinity

Component	Multiply by	To get	Remark
TDS, mg L^{-1}	1.0	TDS, ppm	In soil science, it is now rarely used
Salinity, dS m^{-1}	1.0	Salinity, mmhos cm^{-1}	
Salinity, mmhos cm^{-1}	1000	Salinity, μmhos cm^{-1}	
Salinity in the range of 3 to 30 dS m^{-1}	0.36	Osmotic pressure, atmosphere	
Salinity, dS m^{-1} (< 5 dS m^{-1})	640	TDS, mg L^{-1}	
Salinity, dS m^{-1} (> 5 dS m^{-1})	800	TDS, mg L^{-1}	The factor 800 increases with increasing EC and approaches near unity at very high salinity levels
mg L^{-1}	1/equivalent weight	meq L^{-1}	As an example 80 mg L^{-1} of calcium is equal to approximately 4 meq L^{-1} of calcium, the equivalent weight of calcium being 20.04 mg meq^{-1}

mg L^{-1} = milligrams per litre; meq L^{-1} = milliequivalents per liter; ppm = parts per million; dS m^{-1} = deciSiemens per meter; μS cm^{-1} = microSiemens per centimeter; mmho cm^{-1} = millimhos per centimeter; TDS = total dissolved solids.

The equivalent weight of the element or compound indicates the combining capacity of an element or radical with hydrogen. It is expressed as the weight in grams of the ion or compound that combines or replaces 1 gram of hydrogen. It is evaluated by dividing the atomic weight with valency of the ion or compound. The equivalent weights of some of the ions and compounds are given in Annexure I. The milliequivalent weight is 1/1000 part of the equivalent weight. In the range of 0.1 to 5.0 dS m^{-1}, following relation holds true:

$$1 \text{ meq } L^{-1} = 10 \times EC \text{ (dS } m^{-1}) \tag{2.1}$$

This relation can be used to assess the EC of the water if meq L^{-1} of various cations is known such that:

$$(\text{Sum of the cations in meq } L^{-1})/10 = EC \text{ (dS } m^{-1}) \quad (\text{for EC} < 5.0 \text{ dS } m^{-1}) \tag{2.2}$$

Since sum of anions and cations in the water is equal (within practical limits of measurement), sum of anions can also be used in the above relation.

The EC is easily determined with EC Bridge using soil saturation extract and expressed as EC_e. EC can also be determined using soil : water extracts or solutions such as 1:2 soil water extract/solution (EC_2) or 1:2.5 soil water extract/solution ($EC_{2.5}$). For heavy textured soil even 1:5 (EC_5) soil water extracts/solutions are used. These values then are converted to EC_e for reporting (See chapter 7 and Gupta et al., 2012, Gupta and Gupta, 2014). Although it may be useful to use soil : water extract or paste for quick appraisal of soil salinity problems, caution must be exercised while interpreting the data. The reliability of The EC estimates made on the basis of various soil water ratios depends upon the kind of salts present. For chloride salts, the results may only be slightly affected by moisture content, but for sparingly soluble salts such as sulphate or carbonate, when present in appreciable amount, the results may not be quite accurate and may underestimate salinity. The user's should clearly understand the limitations of the method. Moreover, for the same reason one must report the ratio of soil and water used to determine the EC as it affects the amount and composition of salts extracted. It is customary to report EC at 25°C. Therefore, correction for the temperature is made to the EC value determined from old EC meters. In new meters, this correction is automatically made as the electrode is provided with a temperature sensor.

Table 2.4 EC_e and degree of salinity hazard

Soil salinity class	EC_e (dS m^{-1})	Effect on crop plants
Non-saline	0 - 2	Salinity effect negligible
Slightly saline	2 – 4	Yields of sensitive crops restricted
Moderately saline	4 – 8	Yields of many crops restricted
Strongly saline	8 – 16	Tolerant crops give satisfactory yields
Very strongly saline	> 16	Only a few very tolerant crops yield satisfactorily

Once the soil EC_e is known, degree of hazard can be assessed using the guidelines given in Table 2.4. It helps in crop and varietal selection as well as to devise the cropping pattern in the area.

pH

The water molecules in an aqueous solution get dissociated into H^+ and OH^- ions. The pH of the aqueous solution is log of the inverse of hydrogen ion activity and pOH the log of the inverse of hydroxide ions activity as follows:

$$pH = -\log [H^+] \tag{2.3}$$

$$pOH = -\log [OH^-] \tag{2.4}$$

Kw, the ion-product constant for water is given as

$Kw = [H^+][OH^-] = 1 \times 10^{-14}$ and may be written as:

$$pH + pOH = -\log Kw = 14 \tag{2.5}$$

The pH values on a scale of 0 to 14 characterize the aqueous solutions as acidic, neutral or alkaline. A solution which is neither acidic nor alkaline is neutral and corresponds to a value of 7 on the pH scale. Acidity indicates a higher activity of hydrogen ions showing a pH value below 7. Alkaline solutions characterized by a low hydrogen ion activity show a value above 7. At 0 pH, hydrogen ion activity (mol L^{-1}) is 1.0 being 0.0 at 7.0 and 0.0000001 at pH 14.0. On the other hand, OH^- activity would be 0.00000000000001 at pH 0.0, 0.0 at 7.0 and 1.0 at pH 14.0. Thus, a difference of one pH measurement unit represents a tenfold increase or decrease of hydrogen ion activity in the solution. Based on pH, solutions are characterized as ultra-acids to very strongly alkaline (Table 2.5). In the context of soils and water, an unusually high or low pH is a good indication of presence of some pollutants in the soil/water.

Table 2.5 The descriptive terms for reaction and their respective pH range

Reaction Class	pH range
Ultra acid	1.8 - 3.4
Extremely acid	3.5 - 4.4
Very strongly acid	4.5 - 5.0
Strongly acid	5.1 - 5.5
Moderately acid	5.6 - 6.0
Slightly acid	6.1 - 6.5
Neutral	6.6 - 7.3
Slightly alkaline	7.4 - 7.8
Moderately alkaline	7.9 - 8.4
Strongly alkaline	8.5 - 9.0
Very strongly alkaline	9.1 - 11.0

The pH measurement involves the use of a pH meter having two electrodes, a sensitive (usually glass), and a reference electrode. A temperature element provides a temperature signal to the pH analyzer. The glass bulb is filled with a stable pH solution, usually pH 7. When the outside of the bulb is exposed to a solution or paste where hydrogen ions vary, it develops a potential that is read versus the stable potential of the reference electrode at any given temperature. The reference electrode also completes the pH measuring circuit within the solution. The signal is sent via the silver chloride coated wire to the instrument, where it is measured as a voltage and then converted to pH units. Most pH sensors are designed to produce a 0 mV signal at 7.0 pH, with a (theoretically ideal) slope of -59.16 mV/pH at 25°C. Units that combine both the electrodes into one are supplied with the new meters but the basic principle of measurement remains the same.

Like EC meters, pH meters also require some form of temperature compensation to ensure standardized pH values. Meters with automatic temperature compensation receive a continuous signal from a temperature element and automatically correct the pH for temperature of the solution. Manual temperature compensation requires the user to enter the temperature. pH buffers are also labeled with temperature correction tables. For maximum accuracy and in conditions where the buffers are being brought to *in-situ* temperatures, the tables are used to assess the actual pH.

In a soil water system, the apparent pH values obtained depend on the characteristics of the soil, the concentration of dissolved carbon dioxide, and the moisture content at which the reading is made. Soil characteristics that are known to influence pH readings are: the composition of the exchangeable cations, the nature of the cation-exchange materials, the composition and concentration of soluble salts, and the presence or absence of gypsum and alkaline-earth carbonates.

Like EC, pH_2 and pH_s are also related, pH_2 being higher than pH_s. The difference increases with increasing pH_s, being less in normal and saline than alkali soils (Table 2.6, Gupta, 1972). It may be noted that in alluvial alkali soils of the Indo-Gangetic plains, there is a close relationship between the pH_s, pH_2 and ESP. Unfortunately, this relation does not hold true in many soils. In such cases ESP needs to be determined to calculate gypsum requirement.

Table 2.6 Relationship between pH_s, pH_2 and ESP of the alluvial alkali soils

pH_s	pH_2	ESP
8.0-8.2	8.5-9.2	< 20
8.2-8.4	9.2-9.4	20-35
8.4-8.6	9.4-9.7	35-50
8.6-8.8	9.7-9.8	50-65
8.8-9.0	9.8-9.9	65-75
9.0-9.2	9.9-10.2	> 85

For the salt affected soils of Rajasthan, pH and ESP are found to be related as per the following relation:

$$ESP = -244.77 + 32.23\ pH_2 \quad (2.6)$$

The correlation coefficient (r =0.87) for this equation is highly significant at 1% level. It means that for the soils of this region, soils having pH of about 8.1 will have ESP of 15.

On the basis of pH, the soil is categorized as normal to most of the common crops if pH_2 is 8.5, tending to become alkali (sodic) if pH_2 is between 8.5 and 9.0; and alkali soil, harmful for growing crop if pH_2 is > 9.0; (Muhr *et al.*, 1963). If the pH value of the soil is less than 7.5, it can be presumed that it contains (almost always) no alkaline-earth carbonates.

Soluble Cations and Anions

Complete analysis of soil samples for soluble ions helps to accurately determine the total salt content. Besides, it helps to determine the composition of the salts present, which may also indicate the relative composition of cations on the exchange complex. The commonly determined cations are calcium, magnesium, sodium and potassium and anions are carbonate, bicarbonate, sulphate, and chloride. Sum of the cations should be nearly equal to sum of anions. If sum of cations exceeds the sum of anions after expressing them on an equivalent basis, it may be desirable to determine the nitrate content. It may not be necessary to analyze the sample for other ions such as silicate and boron unless the presence of these is revealed from experience or is required for other studies.

A question often arises as to at which moisture content total soluble ions should be analyzed. Past studies indicate that as the moisture content increases, some ions might be extracted more while others may reduce. Therefore, it is desirable to determine the soluble salts at field moisture content. On the other hand, drawing water extract at such moisture contents might pose few insurmountable problems. Besides, special equipment may be required to draw the extract. Considering all these issues, most analysis work is performed with the saturation extracts.

Cation Exchange Capacity (CEC)

Clay minerals and organic matter constituents of a soil have the cation-exchange properties and are collectively termed as the exchange complex. The dominant residual charge on most soil colloids is negative. Thus, these attract mobile substances of opposite charge i.e. the cations, the positively charged ions in the soil water. Although cation exchange sites immobilize cations yet keep them exchangeable so that these are available to plant roots. Cation exchange capacity has thus been defined as the 'sum total of the exchangeable cations that a soil can adsorb'. The cation exchange capacity is expressed in terms of moles of positive charge per unit mass. For convenience, cation exchange capacity is expressed in centimoles of positive charge per kilogram [cmol (p^+) kg^{-1}] of soil or in meq 100

g^{-1} of soil. The basic principle used to determine CEC is that all the adsorbed cations in the soil are replaced by a common ion, such as barium, potassium, sodium, or ammonium and then the amount of adsorbed barium, potassium, sodium, or ammonium is determined (Richards, 1954, Chapter 7). A rough estimate of CEC can be made by the following equation, which assumes that other cations on the exchange complex are negligible.

$$CEC = Na + Ca + Mg + K \qquad (2.7)$$

To estimate CEC of alkali soils is tedious and complex requiring careful application of an appropriate technique of analysis. Any errors in the estimated CEC will be reflected in the calculated ESP of the soil.

Exchangeable Cations

The relative proportion of cations on the exchange complex is governed by the relative composition of the soil solution. Any change in the composition of the soil solution will also be reflected in the exchangeable cations. But, the total amount of exchangeable cations will remain the same. It may be noted that relative proportions of various cations on the exchange complex govern many physical and chemical properties of soils.

More often, exchangeable cations on the exchange complex are expressed as a percent of the CEC. For example, the soil ESP is given as:

$$ESP = 100\ (\text{Exchangeable sodium})/CEC \qquad (2.8)$$

Herein both the exchangeable sodium and cation exchange capacity are expressed in the same units. On the basis of ESP, degree of alkali hazard of the soil can be assessed as per guidelines reported in Table 2.7.

Table 2.7 ESP and degree of alkali hazard

Approximate ESP	Alkali hazard
Up to 15	None to slight
15 – 30	Slight to moderate
30 – 50	Moderate to high
50 – 70	High to very high
> 70	Extremely high

Sodium Adsorption Ratio (SAR) and ESP

The SAR of the soil solution is calculated from the concentration of cations (meq L^{-1}) in the saturated paste extract by the equation:

$$SAR = Na/\sqrt{(Ca+Mg/2)} \qquad (2.9)$$

If concentrations of the ions are expressed in mmol L^{-1}, then SAR is given as:

$$SAR = Na/\sqrt{(Ca+Mg)} \tag{2.10}$$

Based on eq. (2.10), the units of SAR are (mmol $L^{-1})^{1/2}$ although units based on the basis of eq. (2.9) are also reported.

An equation developed by Gapon (1933) considers that a linear relationship exists between ratios of ions in solution and the same ions on the exchange phase once the ions have been adjusted for valency. Based on the Gapon's equation many equations on cation exchange equilibria appeared in the literature. Perhaps the most widely used application is the one developed by Richards (1954). The relationship between exchangeable sodium ratio (ESR) measured in the exchange phase and sodium adsorption ratio (SAR) measured in the solution phase was reported as:

$$ESR = -0.0126 + 0.01475\ SAR \qquad (r^2 = 0.852) \tag{2.11}$$

The ESR was determined from the exchangeable Na (ExNa) and cation exchange capacity (CEC) by the following relationship:

$$ESR = ExNa/(CEC - ExNa) \tag{2.12}$$

The SAR was calculated from the concentration of cations (meq L^{-1}) in the saturated paste extract by eq. (2.10).

The proportionality coefficient in the regression of ESR on SAR (0.01475 in eq. 2.11) corresponds to the exchange constant of the Gapon's equation. Clearly for a given set of soil properties there exists a unique function between SAR and ESR. Since ESR and ESP are also related, the relation between ESP and SAR is given as (Richards, 1954)

$$ESP = \frac{100(-0.0126+0.01475SAR)}{1+(-0.0126+0.01475SAR)} \tag{2.13}$$

Thus, ESP can be calculated if the SAR of the soil solution is known. If we calculate ESP from eq. (2.13), ESP 15 is obtained at an SAR of 13. It is why SAR 13 has been used as a criterion to diagnose alkali soils.

Later Bower (1959) modified the equation as follows.

$$ESP = \frac{100(0.0057+0.0147SAR)}{1+(0.0057+0.0147SAR)} \tag{2.14}$$

Paliwal and Maliwal (1971) reported that for the Indian arid conditions, following equation performs better than the previously mentioned equations.

$$ESP = \frac{100(0.23+0.0042SAR)}{1+(0.23+0.0042SAR)} \tag{2.15}$$

Similar to the SAR equation, potassium adsorption ratio (PAR) is also calculated to assess K exchange in the soils (Paliwal, 1967). It is not commonly used, probably potassium exchange being much less than Na in alkali soils. Other studies have shown that there is a linear relationship between SAR of the soil solution and ESR up to an ESP of about 50. As such, SAR of the soil solution can be used as a fair measure of the exchangeable sodium status of soils. For a better estimate of exchangeable sodium, the values of Gapon's constant, K_G need to be determined experimentally for each major group of soils.

An unusual aspect of the SAR is that Ca^{++} and Mg^{++} have been lumped together. According to one group of scientists, Mg acts like Ca in influencing the physico-chemical properties of the soil, owing to divalent nature of the cation. Others believe that Mg should not be coupled with Ca in view of the varying affinity for adsorption on the exchange complex of the soil. There seems to be no theoretical basis for combining the concentrations of these two ions, except for the fact that ion valency is considered to be most sensitive parameter for predicting ion-exchange relations as it governs the ions size. Garrels and Christ (1965) have shown that the Debye Huckel distance of the closest approach for magnesium is approximately one-third larger than for calcium. Hence, over the range of solution salt concentration common to irrigation agriculture (2 to 200 meq L^{-1} or 0.001 - 0.1 molar) the activity coefficient for magnesium should be 2 to 20% higher than for calcium. Many soils and clays also tend to prefer Ca^{++} over Mg^{++} in ion-exchange, with a relative increase in preference of about 20% being common. These two effects having opposing trends may account for at least a part of the widespread success for the use of SAR in ion-exchange predictions.

Another reason to combine the Ca and Mg might have been that in the past most analysis results reported the concentration of Ca+Mg together. More important and adequate justification for the widespread use of the SAR is its relative simplicity and the fact that it generally works, although it has not been so under the Indian saline and alkali soils, which may both show high SAR.

In alkali soils calcium and magnesium tend to precipitate resulting in high SAR values sometimes even exceeding 100 because of the sizeable amounts of carbonates/bicarbonates. A relationship between SAR and ESP of the alkali soils of Uttar Pradesh revealed a positive correlation between the SAR and ESP of the soils for all samples having SAR less than 100 (Bhargava, 2003; Fig. 2.1). Out of 41 data sets, 24 data sets had SAR less than 100, which followed a linear relation as per the following equation.

$$ESP = 0.809\ SAR + 12.78 \tag{2.16}$$

Although SAR has been recognized as a parameter to assess alkali conditions defining a nitric horizon as the one having SAR $\geq$ 13, the data from Uttar Pradesh India reveal that for a SAR of 10 one may end up with an ESP of about 21 (Eq. 2.16). Another problem in using SAR as the diagnostic criterion under the Indian condition arises from the fact that even saline soils show high SAR but neither show high ESP nor high pH. It is ascribed to the predominance of chlorides and sulphates and excessive soluble salt concentration of neutral salts that depresses the soil pH.

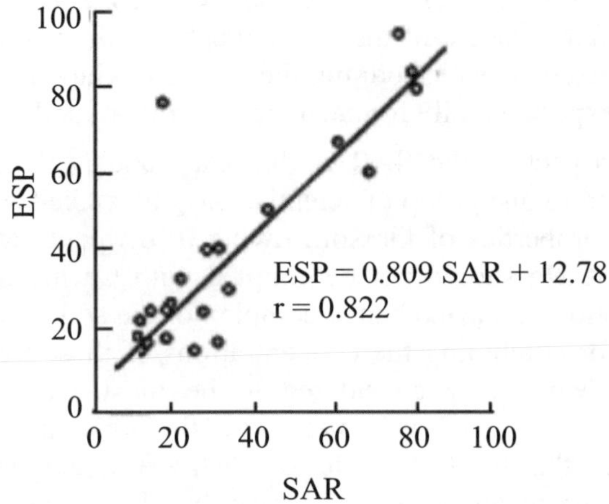

Fig. 2.1 Relationship between SAR and ESP in alkali soils of Uttar Pradesh

Osmotic Effects

Water potential is defined by its chemical activity or free energy. Bound water has less chemical activity or free energy than free water and therefore, has a negative water potential. Water potential is critical for moving water to leaves so that photosynthesis can take place. Total or combined water potential is given as the sum of matric, osmotic, pressure or turgor and gravitational potentials. In saline soils, although the general plant response is related with EC_e, actual plant responses are related to the sum of matric potential and osmotic potential in the root zone. Matric potential is similar to osmotic potential, except that water molecules that are adjacent to a wettable surface, such as a soil particle or a cell wall, are strongly held by hydrogen bonding. This is a very strong cohesive force that holds the first layer of water molecules very tightly, giving them a highly negative matric potential. Additional layers of water molecules are attracted less and less strongly, until the effect becomes negligible. In the soils, this negligible point is considered as the water at field capacity or when the downward movement of water due to gravity stops. Approximately half of this water can be absorbed by plant roots before the matric potential becomes too high for them to overcome with an adequate margin to avoid wilting. The wilting point is usually considered to occur at a soil moisture tension of 15 atmospheres, or a matric potential of about -1.5 MPa. The plant is still obtaining water at this point, but generally not fast enough to meet its needs for transpiration and growth.

Solute potential (Ψ_s), also called osmotic potential, is negative in a plant cell and zero in distilled water. Osmotic potential depends on the concentration of solute ions and molecules in the water. Since an ion has as much effect as a molecule, one mole of an ionic substance such as sodium chloride produces twice as much osmotic potential as one mole of a molecular substance such as sugar. The concentration of some common salts, namely, NaCl, $CaCl_2$ and Na_2SO_4 are

24, 32 and 36 meq L^{-1} per atmosphere (0.101 MPa) respectively. Also, the effect is independent of the size of the molecule or ion, so the amounts required to produce a certain potential are proportional to the ionic and molecular weights of the solute ions and molecules. Since water has lost free energy, so its osmotic potential is assigned a negative sign. The salts which produce large number of ions per molecule and which remain completely dissociated with individual ionic components produce the highest osmotic effect. Furthermore, osmotic effects are additive for all of the solutes that are present in a solution. Therefore, the differences evident for single salt solutions become less pronounced for a mixed salt solution. For mixed salts, the relationship between EC and osmotic pressure (OP) can be related by the following equation.

$$OP = 0.36\ EC = \text{Salinity (mg L}^{-1}) \times 0.000563 \quad (2.17)$$

Here EC is in dSm^{-1}. With a negative sign fixed, OP values in atmosphere become nearly identical with osmotic potential values expressed in bars.

Normally solute potential in cell cytoplasm are -0.5 to -1.0 MPa but it can reach -2, -3, or even more MPa. If the osmotic potential in the soil solution is less than the values in the cell cytoplasm only then the plants will have the driving force to absorb water from the soil and distribute it throughout the plant. Normally the plants growing in non-saline soils are able to draw water as the solutes in the root cells are higher than soil solution. On the other hand in saline soils, osmotic potentials can be quite strong. A total of one mole of solute ions and molecules gives an osmotic potential of approximately -2.5 MPa. Therefore, with increasing salinity of the soil solution, plants are able to draw less water because of the lesser differential between the soil solution and root cells. It is why plants fail to extract water from the soil, even if water is present in the soil.

Matric potential can be as high as zero in a water saturated system but in a dry system, it can be as low as -2 MPa. As the matric potential increases with soil moisture depletion, osmotic pressure also increases proportionally. Since the field capacity moisture content is nearly ½ of the saturation percentage, plants at this moisture content will be exposed to nearly double the values calculated from the soil solution EC_e and four times the values at the lower end of the available moisture content i.e. the wilting point. The data in Table 2.8 clearly brings out this effect that soils are unable to take up the water with increasing soil solution salinity (Kelly and Rengasamy, 2006 as reported in Rengasamy, 2010). Since the adverse impacts of matric and osmotic stress are additive in nature, the only way to minimize the adverse impacts is to keep the matric potential of the soils high. It can be achieved in a number of ways. One way is to apply shallow and frequent irrigations. It helps to manage the matric and osmotic potentials in saline environments. It is why, pressurized irrigation in saline soil or saline water irrigated lands have proven their superiority over surface irrigation methods. The other way is to periodically leach down the salts from the root zone. It can be achieved either by applying fresh water irrigation at critical junctures or by managing the rainfall to leach down the salts.

Table 2.8 Percentage of available soil water not taken up by plants in different soil types due to osmotic pressure of soil water salinity in relation to laboratory measured soil salinity and gravimetric field soil moisture (With permission from Rengasamy)

Laboratory measured soil salinity $EC_{1:5}$ (dS m^{-1})	Gravimetric field soil moisture(%) below which osmotic pressure due to salinity is		Percentage of available soil water not taken up by plants due to osmotic pressure (>1000 kPa) of soil water salinity			
	(>1000 kPa)	(>1500 kPa)	Sand	Sandy loam	Clay loam	Clay
0.25	4.5	3.0	25	0	0	0
0.39	7.0	4.7	50	0	0	0
0.50	9.0	6.0	70	25	0	0
0.72	13.0	8.7	100	50	15	0
1.00	18.0	12.0	100	81	40	4
1.11	20.0	13.3	100	94	50	12
1.25	22.5	15.0	100	100	63	22
1.50	27.0	18.0	100	100	85	40
1.64	29.5	19.7	100	100	98	50
1.64	29.5	19.7	100	100	100	58
2.00	36.0	24.0	100	100	100	76
2.33	42.0	28.0	100	100	100	100

Calcium Carbonate in Soil

The *kankar* or impure $CaCO_3$ is found in many saline or alkali soils developed under arid and semi-arid climates. Calcium precipitates as calcium carbonate nodules or concretions accumulate in the B or C horizon of a soil profile at various depths. These may also accumulate as caliche or cemented hard clay pan of varying hardness. The zone of calcic horizon possibly indicates the zone of shallow water table. As the water table recedes, soil solution is concentrated resulting in precipitation of calcium and magnesium as carbonates. It is a common feature of alkali soils because solubility of calcium is low at high pH exhibited by most alkali soils. For example, solubility reduces from 19.3 meq L^{-1} at a pH of 6.21 to 0.4 meq L^{-1} at a pH of 10.12. Sharma and Bhargava (1981) reported that a distinct, regular and thick zone of calcium carbonate accumulation, mainly dolomitic exists at about 1 m below the surface of alkali soils in the Indo-Gangetic plains. Investigations made at the Usar Reclamation Farm, Chakeri, Kanpur revealed the occurrence of a compact layer 30-40 cm thick and 30-40 cm below the surface, beyond which there is generally a zone of *kankar*, often as a thick pan of 1.2-1.5 m depth. The smooth boundary of the calcic horizon, irregular shape of individual concretions and regularity of occurrence underneath vast areas suggest that these are of pedogenic origin (Sehgal and Stoop, 1972).

Presence of $CaCO_3$ affects both the physical and the chemical properties of a soil. It restricts water movement and prevents root penetration and prolife-ration. A calcareous soil ($CaCO_3$ 2%) usually has pH_s in the range of 7.6-8.4. If $CaCO_3$ is present in the soil profile in moderate amounts, particularly in the root zone depth, it moderates the ESP and prevents excessive deterioration of saline or alkali soils. The rain water owing to its acidic nature converts $CaCO_3$ to $Ca(HCO_3)_2$ having higher solubility than $CaCO_3$. The calcium released from $Ca(HCO_3)_2$ replaces sodium on the exchange complex. While the soluble form of $CaCO_3$ is measured by titration with sulphuric acid and versenate solution, the crystalline and amorphous form of calcium carbonate are measured *in-situ* by its shape, size and release of effervescence on application of dilute hydrochloric acid (Chapter 7). In the presence of $CaCO_3$, it should be possible to use acid or acidifying amendments as well organic materials such as compost, organic garbage, waste acidic products and rejected dairy waste etc. for reclamation. It may be mentioned that most productive normal soils are only slightly calcareous. Highly calcareous soils seriously suffer from nutrient and micro-nutrient deficiencies. The plants grown on such soils sometimes suffer with lime induced chlorosis.

Gypsum

Presence of a small amount (2%) of gypsum ($CaSO_4.2H_2O$) in the soils is considered favourable to crop growth. It helps to maintain soil structure as sodium on the exchange complex is replaced with calcium. Excessive gypsum (25%), on the other hand is harmful to plant growth. Alkali soils of India do not contain sufficient reserves of gypsum. The saline soils, particularly on the Sea coast have been identified to possess gypsum at some depth in the soil profile. Gypsum has also been found to occur in the saline soil profiles of the Rann of Kutch in Gujarat and arid portions of Rajasthan. Satyanarayana (1961) reported that these soils may contain as much as 0.25-1.0% $CaSO_4$.

Toxic Elements

Almost all the essential elements, which are otherwise recognized as micro-nutrients, can prove to be toxic, if present in the soil beyond specific limits. In addition, elements, viz. Al, As, Be, Cl, Cr, F, Pb, Ni, Se and V are also known to cause toxicity to plants, if present beyond permissible limits. The best productive normal soils are generally free of toxic elements or contain less than permissible limits. The saline and alkali soils tend to contain one or more of these elements in toxic amounts. The elements recognized to be of practical importance from the toxicity point of view in saline and alkali soils are B, Li, F, Se and Mo. Out of these five, B, Li and F are phytotoxic. Se and Mo may not adversely affect the plant growth *per se* but forage crops grown on these soils may contain high concentration especially of Mo that may prove toxic to the animals to adversely affect their health. Boron content of 1.0 mg L^{-1} in soils has been considered adequate for the normal growth of most plants. The boron in the saturation extract of normal soils, saline soils and alkali soils may be less than 1.0, 1 to 2 and 2 to 10 mg L^{-1}, respectively (Bhargava *et al.,* 1974). Soluble boron content in alkali soils has been found to increase with pH and ESP as its availability increases at higher pH. On the other hand, high boron in alkali soils may not pose much problem. Boron present as highly soluble sodium metaborate is converted to relatively insoluble calcium metaborate upon reaction with gypsum applied during reclamation (Gupta and Chandra, 1972). Leaching practiced to reclaim saline and alkali soils also helps to reduce the boron content in these soils. Application of gypsum also decreases the Mo in plants due to antagonistic interaction of SO_4 with the adsorption of Mo.

Nutrient Status

The testing of the soil for major nutrients, viz. nitrogen, phosphorus and potassium is a proven practical method to evaluate the fertility status of the soils. These tests provide a sound basis for making recommendations on fertilizer application. The most productive normal soils are high in organic carbon (above 0.75%), which is used as a measure of available nitrogen, high in available P_2O_5 (above 25 kg ha^{-1}) and high in available K_2O (above 280 kg ha^{-1}). A soil is rated as low in organic carbon, available P_2O_5 and available K_2O if the respective values are below 0.5%, 10 kg ha^{-1} and 110 kg ha^{-1}, respectively. The alkali and saline soils are usually low in nitrogen, medium to high in phosphorus and high in potassium. Swarup and Yadav (2004) reported that barren alkali soils usually have high P and K (available P being 34-38 kg ha^{-1}, available K 475-478 kg ha^{-1} in the 0-15 and 15-30 cm layers higher values being in the 0-15 cm layer). Therefore one may not expect much response to the application of P and K fertilizers for few years until the soils are depleted of these nutrients by few crops in succession.

Micro-nutrients

Deficiency symptoms of zinc, iron and manganese have been reported on plants growing on alkali soils. The Zn, Fe, Mn and even Cu content in alkali soils is usually low (Bhumbla and Dhingra, 1964). However, the availability of iron and manganese in alkali soils can be increased substantially by submergence and application of FYM (Swarup, 1981, Swarup, 2000). Since zinc deficiency may still persist, it is useful to apply zinc to crops grown on alkali soils. In saline soils, Cu and Zn content is usually low and Fe and Mn content generally more than the alkali soils.

PHYSICAL PROPERTIES

A soil is a source of water and nutrients for plant growth, is a conduit between soil surface and groundwater and is a filter and a buffer for the pollutants before they percolate to groundwater. The amount of water, air and nutrients available for plant growth on the other hand is greatly affected by the soil physical properties. Apparently, study of soil physics that involves the study of the state and transport of all forms of matter and energy in soils assumes significance. It is more so in salt affected soils especially in alkali soils where soil physical properties are adversely impacted by the presence of sodium on the exchange complex. The status of the soil physical properties is a useful indicator to diagnose the kind and degree of problems of salt affected soils. Besides, these properties help to decide the way these soils should be effectively managed. The soil physical properties that have practical implications for salt affected soils are listed in Table 2.2, some of which are described in the following sections.

Effective Soil Depth

The soil depth that can be effectively exploited by plant roots is an important element for good plant growth. Excellent yields are obtained on normal soils having an effective root zone depth varying from 45 to 90 cm. Shallow-rooted crops do fairly well even if soil depth is just 30 cm. High ESP of the alkali soils limits the effective soil depth to 10-15 cm. As such, root penetration is seriously inhibited in alkali soils and may strongly influence the plant growth in some deep rooted crops such as wheat, gram, mustard and cotton. Besides, the presence of an impermeable *kankar* layer also affects the soil depth and may prove to be a major hindrance in the cultivation of deep rooted crops and plants. Minor improvements can be brought through application of higher doses of amendments but overall these observations imply that the zone of soil moisture extraction in alkali soils is limited. Though few crops adapt to edaphic environment by modifying their root system, it is useful to grow only shallow rooted crops at least in the early phase of reclamation. In saline soils, such restrictions may not be encountered except in cases where *kankar* and gypsum accumulation is at shallow depth.

Bulk Density and Porosity

Bulk density is defined as the dry weight of a unit volume of soil and is expressed in $g.cm^{-3}$ or $Mg.m^{-3}$. The bulk density of highly productive normal soils usually range from 1.1 to 1.5 (fine texture to medium texture) and 1.2 to 1.65 g cm^{-3} (coarse texture). It may go up to 1.8 g cm^{-3} for sands. The alkali and saline soils (medium texture) may have bulk density in the range of 1.5 to 1.65 g cm^{-3}. Because of dispersion effect of sodium, bulk density in top surface layer of alkali soils can go up to 1.7 g cm^{-3}. High bulk density may impede drainage and inhibit root penetration and proliferation. In medium or fine textured soils, the infiltration rate and permeability are adversely impacted when bulk density exceeds 1.65 g cm^{-3}. Thus, high bulk density may become a critical factor in achieving high productivity. The bulk density of alkali soils tends to decrease with application of amendments, tillage and cropping.

Pore space is closely related to bulk density of the soils. Soils having high bulk density will have less pore space than soils having low bulk density. Soils with a low porosity generally contain little air space at field capacity. In highly productive normal, medium to fine textured soils, the non-capillary porosity in the plough-zone is usually 10-30%. On the other hand non-capillary porosity as low as 2% has been reported in alkali soils adversely affecting the normal aeration. Gupta and Narain (1971) observed that the total porosity and non-capillary porosity in normal soil profile (0-110 cm) varied from 45.8 to 38.5% and 8.6 to 5.7% as compared to 39.5 to 38.9% and 4.9 to 2.5% in saline-alkali soils.

The particle density of many soils is averaged around 2.65 g cm^{-3}. Measurement procedures for bulk and particle density are discussed in Chapter 8 of this handbook. The porosity of soil (n) may be obtained directly from air-pycnometer. It can also be calculated from the following relation

$$\text{Porosity, } n = 1 - BD/PD \qquad (2.18)$$

Here BD and PD are the bulk and particle densities.

Infiltration Rate

Infiltration rate is defined as the rate at which a soil, in a given condition at a given time, can absorb water. It is also the rate at which a soil will absorb water ponded on the surface as a thin sheet when the ponded area is infinitely large or when adequate precautions are taken to minimize the effect of horizontal flow at the borders. Although infiltration is treated as a property of the surface layer of the soil, but when the rate of water entry through the soil surface is not limiting, then the infiltration rate is governed by the downward movement through the profile. Infiltration rate is the volume of water passing from the soil surface into the soil per unit area per unit time. The infiltration rate has the dimensions of velocity, L/T, where L is the length and T the time, commonly expressed in cm h^{-1}. The initial intake rate rapidly decreases as the soil is wetted. The asymptotically decreasing curve has a tendency to approach a final minimum value designated as the basic or equilibrium infiltration rate. It may be attained in few to several hours depending upon the soil type. Both the initial infiltration rate and the basic infiltration rate decrease with the increase in ESP (Table 2.9). Similar results have been reported by Tiwari *et al.* (1981). The most productive normal soils have basic infiltration rates between 1.0 and 3.5 cm h^{-1}. The alkali soils generally have basic infiltration rate below 0.1 cm h^{-1} characterizing the soils as non-arable except for rice crop. In general, the infiltration rate of alkali soils is relatively low than the adjoining normal soils being as low as 1/100 in extreme cases (ESP >70). In a sodic Vertisol, cumulative infiltration decreased sharply with increase in ESP up to 15 and was almost negligible at an ESP of >35 (Table 2.10). If the cumulative infiltration rate is described by Kostiakov's power function relation (cumulative infiltration, $I = at^b$), then the both the constants, a and b of the power function change with increasing ESP. While exponent b decreased with increasing ESP both in alluvial and black sodic soils, constant a increased in alluvial (Acharya and Abrol, 1991) while remained almost same in black sodic soils (Ranade and Gupta, 1987). The data on cumulative intake reveals that only limited amount of water per irrigation need to be applied to alkali lands even after gypsum application so as to avoid water stagnation (Table 2.11). It also point out towards problems that may be encountered in leaching and in irrigation water management of alkali lands even after few years of land reclamation. Infiltration rate of saline soils on the other hand may not be much different than the adjoining normal lands.

Table 2.9 Infiltration rate as affected by ESP

Soil ESP	Initial infiltration rate ($cm\ hr^{-1}$)	Basic infiltration rate ($cm\ hr^{-1}$)
5	22.1	3.7
14	20.3	2.7
22	11.9	1.4
30	9.9	1.0
44	8.6	0.7
49	5.9	0.3
62	5.4	0.2
65	2.2	0.1
70	2.0	0.06
72	1.8	0.05
74	1.7	0.04

Source : Singh (1980)

Table 2.10 Infiltration rate at various ESP levels in an alkali Vertisol

ESP	Kostiakov equation for infiltration rate	Basic infiltration rate ($mm\ hr^{-1}$)
10	$I = 5.7\ t^{0.5}$	4
12	$I = 5.7\ t^{0.36}$	1
15	$I = 4.9\ t^{0.31}$	0.5
>35	Approached to almost 0	Approached to almost 0

The principal methods used to measure infiltration rate are:

1. Rainfall simulators
2. Runoff plots
3. Basin method
4. Furrow method
5. Cylinder (ring) infiltrometers

While basin and furrow methods are commonly used in irrigation engineering, rainfall simulators and runoff plots are used in hydrology and soil conservation. Cylinder infiltrometer method is most commonly used method in irrigation and drainage engineering (Chapter 8). The length of time a test is conducted should be enough to arrive at the basic infiltration rate. Besides, enough water must pass through the soil as may be required to leach the salts in a salt affected soil. Another point that must be kept in view is that the water quality used to conduct infiltration tests should be the same as will be used for irrigation or leaching. Otherwise, one may end up with misleading conclusions.

Table 2.11 Cumulative water intake (cm) in a normal and an alkali land amended with different quantities of gypsum

Soil	Gypsum added ($t\ ha^{-1}$)	Time (hours)			
		1	10	15	20
Normal	-	2.4	7.4	9.1	10.4
Alkali	0	0.6	1.9	2.4	2.8
	7.5	0.8	2.6	3.2	3.7
	15.0	1.1	3.4	4.2	4.8
	22.5	1.2	4.1	5.1	6.0
	30.0	1.2	4.2	5.3	6.2

Hydraulic Conductivity

Permeability is a qualitatively term used to describe the quality or state of a porous medium relating to its readiness with which it conducts or transmits fluids. When defined quantitatively, the term hydraulic conductivity is used. It is defined as the constant of proportionality between the flux and the total driving force in Darcy's law or as the flux caused by a unit driving force. Several factors, viz. texture, structure and structural stability, consistency, $CaCO_3$, cleavage planes, visible pores and depth to impermeable strata such as bedrock or a hardpan influence the hydraulic conductivity of the soils. The most productive normal soils have reasonably good hydraulic conductivity. It ensures that saturated conditions, whether from rainfall, irrigation or both, are unlikely to occur for more than 48 hr in the upper root zone. Hydraulic conductivity of saline soils is also good because excess salts keep the clay in flocculated state. On the other hand alkali soils usually show poor hydraulic conductivity because high exchangeable sodium keeps the clay particles in dispersed state.

The hydraulic conductivity in alkali soils decreases with increasing ESP of the soils. Abrol et al. (1978) reported that while an increase in ESP from 5 to 14 resulted in slight decrease in saturated hydraulic conductivity, further increase in ESP brought about a sharp decline in alluvial alkali soils as revealed by relative hydraulic conductivity (RHC) (Fig. 2.2). The decline is much sharper in Vertisols with RHC becoming almost zero at ESP >35 (Fig. 2.2). The decrease in basic infiltration rate reported in the previous section, which can also be taken as the vertical hydraulic conductivity under certain conditions, also attests to the statement that hydraulic conductivity reduces with increasing ESP.

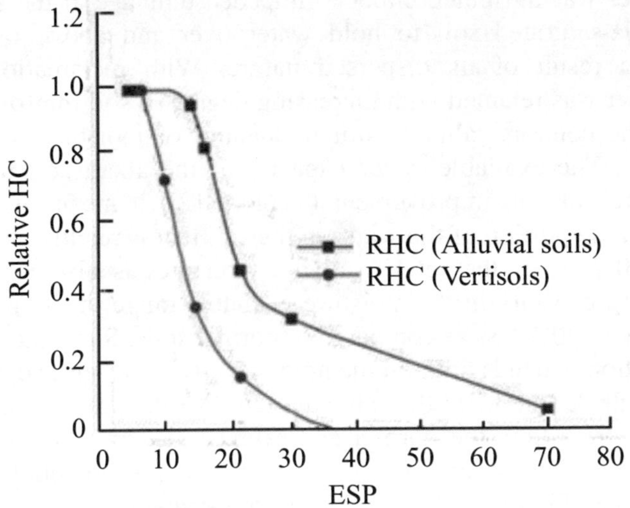

Fig. 2.2 Relationship between saturated hydraulic conductivity and soil ESP

A number of methods are described in the literature to measure saturated hydraulic conductivity. The principal methods used to estimate/measure the hydraulic conductivity are:

1. Correlation methods (relation of hydraulic conductivity with soil properties such as pore size and particle size (grain size) distributions, and soil texture
2. Laboratory methods namely constant and variable head methods (Chapter 8)
3. Field methods such as auger hole and inverse auger hole methods (Chapter 8)
4. Cylinder (ring) infiltrometers

Like saturated hydraulic conductivity, unsaturated hydraulic conductivity is also affected by soil ESP. The unsaturated hydraulic conductivity at 100 cm suction has been observed to be 0.05 cm d^{-1} in an alkali as compared to 0.6 cm d^{-1} in a normal soil (Abrol and Acharya, 1975, Acharya and Abrol, 1975). The difference in hydraulic conductivity of two soils decreased further with increasing suction.

Soil Water Availability

The capacity of a soil to retain water available to plants has a direct bearing on the depth of irrigation water application and its frequency. Water retention generally decreased with increasing ESP. Acharya and Abrol (1975) observed that following prolonged infiltration and subsequent drainage for 48 h, the alkali soil in the subsurface layers held less water; 5 to 6% by weight or 8 to 9% by volume, in comparison to the normal soil. Somewhat higher moisture retained in the surface layer was attributed either to impeded drainage at the surface or the tendency of Na-saturated soil to hold water over and above the pore-space saturation as a result of its dispersed nature. With reclamation, increasing quantity of water was retained with increasing degree of soil improvement at low suctions. On the contrary, almost similar quantity of moisture was retained at higher suctions. The available water capacity of the alkali soil increases with increasing degree of soil improvement (Table 2.12). It means that the readily available water, the portion of the water in the soil that is readily absorbed by the roots (about 50-75% of the total available moisture) is also increased. The moisture holding capacity in the moisture available range of the plants in alkali soils may be about 20% less as compared to normal soils. Since the concentration of the soil solution depends both on the amount of soluble salt and the amount of water present, alkali soils with the same amount of salts present may experience higher salinity stress than the normal and saline soils. In saline soils, moisture availability to the plants decreases due to high osmotic potential of soil water solution as discussed in one of the sections of this chapter.

Table 2.12. Soil moisture retention characteristics of an alkali and gypsum treated soil

Soil moisture tension (cm)	Soil water content ($cm^3 cm^{-3}$)	
	Gypsum treated soil	Alkali soil
50	0.35	0.27
100	0.26	0.24
150	0.24	0.22
200	0.22	0.21
300	0.21	0.20
400	0.20	0.19

Dispersion Coefficient

The forces that bind clay particles together are disrupted when too many large sodium ions come between them. When this separation occurs, the clay particles expand, causing swelling and soil dispersion. Soil dispersion causes clay particles to move and plug soil pores. It results in reduced soil permeability. When soil is repeatedly wetted and dried, dispersed clay particles solidify into almost cement-like soil with little or no structure.

Dispersion coefficient defined as the ratio of water dispersible clay to the total clay expressed as percent is usually very high in alkali soils. It also increases with increasing ESP (Fig. 2.3). The dispersion coefficient exceeds 90% in soils having pH greater than 10.0 compared to 20 to 30% in normal soils. The soils with illitic clay are more sensitive to dispersion and clay movement than those with montmorillonite clay. The difference in the flocculation value is probably due to a smaller attraction force in Na-illite. Na-illite particles have an average thickness of about 10 mm and the planer surfaces are terraced. Once the particles approach closer to each other, the unavoidable mismatch of the terraces leads to poor contact between the edges and the surfaces leading to smaller edge-to-face attraction forces and consequently, a higher flocculation value for Na-illite than Na-montmorillonite (Oster, 1982). Unlike the alluvial soils, the dispersion ratio in Vertisols increases steadily up to 15 ESP but thereafter remains around that level up to an ESP of 60. This suggests that the flock value in Vertisols at 15 ESP is such that the smectitic alkali soil presents similar structural degradation at ESP 15 and ESP 60 because of high buffering nature of clay mineral. High dispersion coefficient is responsible for the following commonly observed features in alkali soils.

- Reduced infiltration and hydraulic conductivity due to sealing of surface pores under the impact of rainfall or when a water of low salinity is applied
- Higher sediment content in the run-off waters from alkali catchment
- Surface crusting and formation of a layer of high clay content in the profile at some depth.

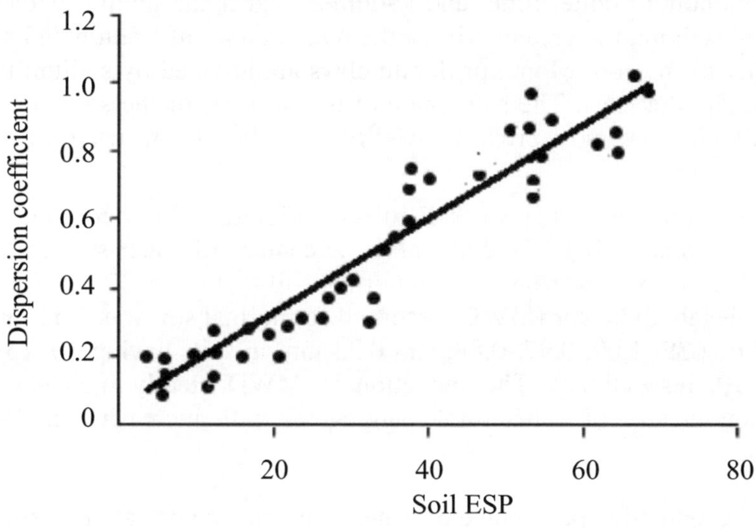

Fig. 2.3 Effect of soil ESP on dispersion coefficient

Clay Swelling

Soil colloids owing to their small size, large surface area, and net negative charge have extremely reactive surfaces. As a result, the clay content significantly influences the soil hydraulic properties. Exchangeable cations are constrained within the electrical influence of the negatively charged clay particles, which are attracted to the charged surface when their concentration is high into the bulk solution. As a result, clay particles act as miniature osmometers and imbibe water to lower the ion concentration near charged surfaces. The action is referred to as swelling. Sodium montmorillonite swells freely. The free swell in black cotton soil is due to the high clay content and presence of swelling mineral Volkonskoite. Large swelling pressures develop between sodium montmorillonite platelets, and single platelets tend to persist in dilute solutions. Increasing the electrolyte concentration on the other hand decreases swelling. It is because many ions in the soil solution counteract the swelling of the soil so that saline soils usually do not have unfavorable physical properties. Divalent calcium ions are more strongly adsorbed to the clay to swell. Individual calcium montmorillonite platelets tend to aggregate into packets or tactoids of several clay platelets with a 0.45 mm film of water on each internal surface.

Aggregation and Stability of Structure

The arrangement of soil particles into crumbs or aggregates that are more or less water stable is an important aspect of soil structure. Alkali soils often have a dense, blocky, single-grain structure, are hard to till when dry, and have low hydraulic conductivity when wet. This is generally because the aggregates and also the pores of such soils are not stable. The aggregates slake down in water, and the pores are filled with fine particles. Although sodium binds to all clay types i.e. montmorillonite, illite, and kaolinite, higher the ability of the type of clay to hold sodium, the greater will be the reduction in infiltration and hydraulic conductivity of the soil. Montmorillonite clays are affected by sodium the most, while kaolinite the least. The same pattern is also true for the swelling of soils. Montmorillonites are most prone to swelling and dispersion, whereas kaolinites are the least prone.

The soil structure of black alkali soils deteriorates with increasing soil ESP (Singh and Verma, 2016). The cumulative percentage of water-stable aggregates of 1.0 mm size decreased from 61.0 in normal soil to 4.0 at ESP 58 (Table 2.13). The mean weight diameter (MWD) recorded for normal soil was 1.52 mm which decreased to 1.39, 1.09, 0.97, 0.88, and 0.23 mm in soils having 10, 15, 22, 38, and 58 ESP, respectively. The reduction in MWD clearly indicates a sharp deterioration in size of water-stable aggregates with increasing exchangeable sodium content.

Table 2.13 Cumulative percentage of water stable aggregates at different ESP

ESP	Water-stable aggregates (mm)					
	>2	>1	>0.4	>0.2	>0.08	>0.05
6	40.0	61.0	77.8	88.2	92.0	92.7
10	35.8	54.8	74.0	83.2	86.6	88.3
15	20.6	44.6	70.7	83.9	90.1	92.0
22	18.2	36.0	63.6	79.1	84.7	86.2
38	18.0	38.3	49.7	64.6	71.1	74.2
58	0.8	4.0	19.4	34.3	45.	51.4

Many studies have reported the potential of soil conditioners to improve soil structure of alkali and saline soils. Although, not many studies have been conducted in India, yet the limited studies have revealed that the conditioners are costly and have limited benefits. Moreover, the benefits are short lived. In the absence of chemical amendments no perceptible improvements have been obtained. River sand either mixed or placed at the soil surface @ 9 and 18 kg m^{-2} resulted in initial increase in infiltration rate but the benefits got neutralized and there was no perceptible improvement in water intake (Acharya and Abrol, 1976). After 168 h of the test, cumulative infiltration was 10.86, 11.46 and 10.72 cm in case of control and mixing of sand at the two rates of 9 and 18 kg m^{-2}. It seems that further investigations need to be initiated on this issue. Organic-matter alone on the other hand proved beneficial albeit very slowly. However, synergy of chemical amendments and organic matter has proved the effectiveness of organic matter in hastening the land reclamation process.

Crust Formation

Soils that have low stability of structure disperse and slake when they are wetted by rain or irrigation water and as the soil surface dries may develop a hard crust. This crust presents a serious barrier to the emergence of seedlings and with some crops it could be the main cause of poor stand. Although normal or saline soils may also develop crust, yet alkali soils are most prone to crust formation, having low stability of structure. Hard surface crusts are typically associated with soils having high ESP and low organic matter content, the two main characteristics of alkali soils. Puddling and wetting the soil to zero tension may also enhance crust formation. Typically alkali soil crust having a total salt content of 22.8% was found to contain 48.3% Na_2CO_3, 33.4% $NaHCO_3$, 13.3% Na_2SO_4 and 3.9% NaCl.

Modulus of rupture, a property that is used to infer the strength of the crust, is found to increase with increasing ESP of the soil as per the following relation (Sandhu et al., 1980):

$$Y = 3.92 + 0.204 \text{ ESP} \qquad (r = 0.92) \qquad (2.19)$$

Here, Y is the modulus of rupture in bars.

Cracking Behaviour

Due to high amounts of sodium on the exchange complex, the soils are highly dispersed and upon wetting, they get puddled and become hard when drying. Thin flakes get peeled off as the drying continues. If drying continues, large cracks are formed in alluvial soils of the Indo-Gangetic plains, which enhance the loss of water from the deeper soil layers. On the contrary, black alkali soils (Vertisols) do not produce large and wide cracks at higher ESP. In contrast to the qualifying character of these soils having deep wide crack, alkali Vertisols at high ESP exhibit narrow, shallow, and higher numbers of cracks.

BIOLOGICAL PROPERTIES

Organic Matter

Soil organic matter abbreviated as SOM or simply the organic matter, OM is also referred as humus. It is found in small amounts but has a great bearing on the productive potential of agricultural soils. Soil organic matter or humus comprises of non-humic and humic substances. While the non-humic substances are the unaltered remains of plant and animal tissues including the organic compounds, the humic substances are chemically and biologically modified substances, being the most active fraction of the organic matter. Humic substances are resistant to microbial degradation. Based on their solubility, these are separated into four fractions namely (i) fulvic acid soluble in both alkali and acid (ii) humic acid, soluble in alkali but insoluble in acid and alcohol; (iii) hymatomelanic acid, soluble in alkali and alcohol and (iv) humin, insoluble in both alkali and acid. Humus found in small amounts in productive agricultural soils; generally representing less than 5% of the dry weight, imparts unique and remarkable properties to the inorganic matrix of sand, silt and clay. In its various forms, it greatly impacts the physical, biological and chemical properties of the soil. OM acts as a long-term carbon sink, and as a slow-release pool for nutrients. It contributes to ion exchange capacity (nutrient storage), nutrient cycling, soil aggregation, and water holding capacity, and it provides nutrients and energy to the plant and soil microbial communities. Soils with high organic matter tend to require lower farm inputs, and to be more resilient to drought and extreme rainfall events. The organic matter content in saline and alkali soils of arid and semi-arid regions in their virgin state is usually low. In alkali soils, the alkaline solutions of sodium salts being solvents for soil humus, the latter is highly dispersed. On evaporation, a dark coloured residue is left in the form of a thin crust or rings around small depressions, thereby imparting a dark black colour to the surface of the soil. The OM content in alkali soils is, therefore, very low, generally in the range of 0.1-0.5%. Evidences have been generated to show that organic matter tends to counteract the unfavorable effects of exchangeable sodium on soils. It is why, one of the objectives of improved cultural practices adopted during reclamation and cropping is to increase the OM content of the soil. Cultivation of green manure crops in the initial years of reclamation is one such practice recommended by the CSSRI. There are hundreds of millions of micro-organisms such as bacteria, actinomycetes, fungi, algae and protozoa in a spoonful of soil, Most are beneficial to soils and the plants growing on them. They form organic bonds that hold soil particles together and thus, help maintain soil structure. They also release carbon dioxide, energy and mineral nutrients that plants use for their growth. When they die and get decomposed, they release more nutrients for plant use.

The physical and chemical properties of a soil determine the nature of the environment in which micro-organisms are found. Singh (1950) listed the species of blue-green algae found in sodic soils of northern India. Saxena (1960) collected the algal flora of sodic soils and observed that if pH is high (9.0-10.9) and the total nitrogen content of the soil is low (below 0.2%), the population of algal flora is likely to be poor. In ditches with stagnant water, where pH varied from 9.0 to 10.2 and total nitrogen content was about 0.03%, green and blue algae were in equal proportions. As the pH was lowered (7.9-8.6), the proportion of green algae increased. The development of the algal flora is the most prominent during monsoon. The total number of bacteria, actinomycetes and fungi are usually the highest in productive normal soil followed by saline and sodic soils (Table 2.14). These environmental characteristics in turn affect the composition of the microbial population both qualitatively and quantitatively. Total organic carbon (%), dehydrogenase activity (μg TPF g^{-1} soil), microbial biomass carbon (mg kg^{-1} soil), % microbial biomass carbon and DHA/Misc, all were low in barren alkali soils. (pH = 10.6, ESP = 95). The values were 0.5 4.5, 56.7, 2.3 and 0.08 respectively (Batra *et al.,* 1997). All values improved to 0.44, 112.1, 203.3, 4.6, 30.55 upon reclamation with Karnal grass and gypsum applied at the rate of 50% GR (pH - 9.3, ESP = 40).

Table 2.14 Microbial counts in different soils

Nature of the soil	Organisms (c.f.u× 10^5 g^{-1} soil)		
	Bacteria	Actinomycetes	Fungi
Normal	96.00	11.6	0.21
Saline (EC_2 6 dS m^{-1})	11.30	0.4	0.045
Sodic (pH_2 10)	2.33	0.04	0.0003

Source : Neelkantan and Pahwa (1977)

FACTORS AFFECTING PLANT GROWTH IN

Several factors come into play to adversely impact crop growth when crops are cultivated on salt affected soils. Separate list of factors for alkali and saline soils are given in this section. Usually, a combination of the factors is in operation to affect crop production and productivity in salt affected soils.

Alkali Soils

Plant growth, yield and/or quality of crops in alkali soils are mainly affected by the adverse physical and chemical make-up of the root zone. The degree of damage depends upon the amount of exchangeable sodium, pH, nature and kind of crops grown, macro and micro-environmental conditions and management interventions (Gupta *et al.*, 2002).

- High ESP and pH due to high concentrations of HCO_3 and CO_3
- Poor water and air permeability as a result of high dispersion of soil aggregates and clay particles
- Toxicity of Na, HCO_3 and CO_3 ions

- Excess of Na causing imbalance in the uptake of other ions such as K and Ca nutrition due to antagonistic effect
- Decreased solubility and availability of micro-nutrients like Zn, Fe and Mn due to high pH and presence of $CaCO_3$ and soluble CO_3 and HCO_3 ions
- Presence of hard $CaCO_3$ layer at or around 1.0-2.0 m depth, which act as a physical barrier for the movement of water, salts and roots of crops and trees
- Hard crust on the surface layer which greatly hinders the seedling emergence and reduces the germination percentage and plant population
- Poor drainage causing aeration problems
- Low organic matter content.

Saline Soils

In saline soils plant growth is adversely affected due to the following factors.

- Excess soluble salts present in the soil solution depress the water potential of the medium that restricts the water uptake by plant roots. It is often referred as physiological drought
- High concentrations of some specific ions (e.g. Na^+, Cl^-, SO_4^{--}, Mg^{2+}, B) can be toxic and can induce physiological disorders
- High salinity may severely affect several metabolic processes such as CO_2 assimilation, protein synthesis and respiration
- In waterlogged salt affected soils, some elements such as aluminium, iron and manganese may be produced and can accumulate to cause toxicity
- Imbalance in the uptake of ions.

SOIL PROFILE DESCRIPTION

Chemical analysis of an alkali and a saline soil profile is given in Table 2.15 a, b and Table 2.16. The data are used to highlight the differences in the chemical characteristics of the two kinds of soils and to illustrate the interpretation of chemical analysis results.

Alkali Soil

The Zarifa Viran series is a member of the fine-loamy, mixed, hyperthermic family of *Typic Natrustalfs*. Zarifa Viran soils have pale yellow to yellowish brown very strongly alkaline loam to clay loam A horizons, olive brown to light olive brown very strongly alkaline loamy C horizons (Table 2.15 a, b). They have developed in old alluvium in micro depressions on nearly level to flat lands at an elevation of 200 to 250 m above MSL in Karnal district, Haryana and in parts of

Uttar Pradesh. The climate is semi-arid sub-tropical with mean annual air temperature of 25°C and mean annual rainfall of 600 to 700 mm.

The pH (1:2.5) readings are greater than 8.5 and are as high as 10.3 in the top 24 cm. pH decreases with depth but continues to be high, in the alkali range. High pH indicates the presence of CO_3, which is present in sufficient quantity. Sum of CO_3 and HCO_3 is higher than sum of the Cl and SO_4. EC_e is high in the top 24 cm but is less than 2.0 in the lower profile. Typically the soil can be characterized saline-alkali but considering the low salinity in the profile and need for amendment application to effect leaching, soil is designated as alkali. ESP is high throughout the profile being as high as 97. Nothing can grow on this soil until exchangeable sodium is brought down to the permissible level for the crop to be grown. $CaCO_3$ is present and increases with increasing depth. Thus, acid forming amendments can be applied to reclaim these soils.

Saline Soils

The saline soil profile reported in Table 2.16 is from District Jind in the state of Haryana. The EC_e is more than 4.0 dS m^{-1} and is nearly similar in the upper 60 cm. The increase beyond 75 cm depth is very apparent (Table 2.16). The pH of the saturated soil paste is less than 8.5 throughout the profile. ESP is also low, being less than 10 throughout the soil profile. On the basis of these characteristics, the soil can be categorized as saline soil. Chloride and sulphate are the principal soluble anions. Bicarbonate content is relatively low, and carbonate is absent. SAR (not calculated) is low throughout the profile. There is no need to add any amendment to reclaim this kind of soil. Boron is high throughout the soil profile and may require additional leaching water to bring it within the permissible limit.

Table 2.15a Physico-chemical characteristics of Zarifa viran series (*Typic Natrustaf*)

Horizon	Depth (cm)	Size class			Coarse fragments >2 mm, % of whole soil	Organic carbon (%)	Carbonate as $CaCO_3$< 2mm (%)	pH (1:2.5 soil water suspension)	Bulk density (g cm^{-3})	EC_e (dS m^{-1})	CEC/ Clay ratio
		Sand (2-0.05 mm)	Silt (0.05-0.002 mm)	Clay (<0.002 mm)							
A11	0-5	43.4	34.6	22.0	-	0.30	0.5	10.3	-	8.2	0.46
A12	5-24	33.4	37.6	29.0	-	0.30	0.9	10.3	1.48	8.0	0.44
B21t	24-56	30.6	36.2	33.2	-	0.20	1.4	9.8	-	1.9	0.45
B22t	56-85	26.0	42.6	31.0	4	0.20	3.3	9.8	1.63	1.4	0.46
B3ca	85-118	33.2	40.0	26.8	10	0.10	12.4	9.6	1.47	1.0	0.41
Cca	118-140	45.0	32.0	23.0	-	0.10	20.5	9.2	-	0.9	0.41

Table 2.15b Physico-chemical characteristics of Zarifa viran series (Typic Natrustaf)

Depth (cm)	Exchangeable bases (c mol (p^+) kg^{-1})					Saturation moisture content (%)	Ionic composition of soil saturation paste extract (mmol L^{-1})							DTPA extractable micronutrients (mg L^{-1})			
	Ca+Mg	Na	K	ESP	CEC NH_4OAC		Ca+Mg	Na	K	CO_3	HCO_3	Cl	SO_4	Zn	Cu	Mn	Fe
0-5	0.2	9.9	0.1	97	10.2	33	0.4	85.3	0.2	30.0	36.5	12.2	5.5	0.36	2.01	12	9
5-24	0.3	12.1	0.4	94	12.8	39	0.4	83.9	0.1	27.0	40.5	14.5	5.5	0.36	2.32	9	16
24-56	0.8	13.4	0.6	90	14.8	32	0.6	18.0	0.1	2.5	8.5	5.5	3.0	0.34	1.38	11	6
56-85	1.8	12.4	0.4	85	14.6	40	0.5	13.5	0.1	2.2	7.6	2.5	2.0	0.21	0.80	8	5
85-118	3.1	7.6	0.5	68	11.2	40	0.8	10.0	0.1	2.2	4.5	2.5	1.0	0.21	0.50	6	4
118-140	5.6	3.8	0.4	39	9.8	44	1.0	9.2	0.1	1.5	5.0	2.5	1.0	-	-	-	-

Clay fraction mineralogy is MI4 KK3 M3; MI=mica, KK=Kaolinite and SM= Smectite (4 means half to one third, 3 means one third to one fifth and 2 means one fifth to one twentieth

Table 2.16 Physico-chemical characteristics of a saline soil from Jind district (Haryana)

Depth (cm)	pH_s	EC_e (dS m^{-1})	Ionic composition of soil saturation paste extract (mmol L^{-1})							Boron (mg L^{-1})
			Ca+Mg	Na	K	CO_3	HCO_3	Cl	SO_4	
0-11	7.5	4.5	20.5	35.6	0.6	Nil	1.0	35.5	21.5	11
11-30	7.6	4.6	30.7	22.6	0.5	Nil	1.0	40.5	10.0	6
30-56	7.8	4.8	32.6	23.6	0.3	Nil	1.0	37.5	12.7	6
56-76	7.8	5.6	30.5	27.3	0.2	Nil	1.0	39.5	14.5	7
76-141	8.0	8.3	41.5	42.8	0.2	Nil	6.0	58.0	22.5	10
141-200	8.0	12.7	50.7	74.1	0.2	Nil	10.0	70.5	39.5	14

Depth (cm)	Sand	Silt	Clay	$CaCO_3$ (%)		CEC (cmol(p+) kg^{-1})	OM (%)	ESP
		(%)		<2 mm	>2mm			
0-11	69.3	21.2	9.5	Nil	-	4.4	0.5	6.8
11-30	61.5	24.9	13.6	0.8	-	5.7	0.3	7.0
30-56	60.9	21.8	17.3	2.9	-	7.8	0.3	7.6
56-76	55.7	21.8	22.5	4.4	7.0	8.9	0.2	8.9
76-141	57.0	21.6	21.4	8.0	35.0	8.9	0.2	8.9
141-200	56.1	22.0	21.9	16.8	15.0	7.6	0.1	9.2

Chapter - 3

RECLAMATION AND MANAGEMENT OF SALINE SOILS

Saline soils are a part of the landscape of arid, semi-arid and coastal ecologies. A glance at the distribution of saline soils suggests that the problem of soil salinity can be categorized under the following three groups.

- Waterlogged saline soils of the irrigation commands
- Saline soils formed as a result of irrigation with saline groundwater
- Coastal saline soils

Processes of saline soil formation under these categories and ecologies served by canal irrigation and groundwater differ widely. Similarly, saline soils formation in coastal ecologies has some additional elements that accentuate salinity problem. The factors causing soil salinization in all the regions get stimulated upon introduction of irrigation through canal or groundwater. As an example overexploitation of groundwater in coastal regions initiates sea water intrusion and is one of the major causes of worry in the soil salinization scenario around the globe. Although the factors affecting soil salinization may be different but basic principles of reclamation and management of saline soils under the three scenarios are similar, although few management practices may differ from one scenario to another. It must be realized that saline soils may not ever be fully reclaimed in the true sense of the word reclamation. The reclamation processes in these soils only brings the salts in the root zone to an acceptable level and thereafter, the land needs to be managed to keep the salts within the permissible limits of crops grown. Thus, living with the salts is an important aspect of a saline land reclamation and management programme. This chapter discusses the basic principles of reclamation and management of saline soils (Table 3.1). Wherever required, region or cause specific management practices are included.

SCRAPING: A WASTEFUL EXERCISE

In order to get quick relief, farmers scrape the top few cm of the soil which contains the maximum salts in areas having shallow water table. Scientifically speaking, the practice may provide some temporary relief for a period of 1-2 years. On the contrary, it has limited utility because the reduced depth to water table hastens the process of capillary action, increases the evaporation rate and results in quicker soil salinization (Fig. 3.1). Besides, disposal of scraped salty soil also poses a major problem. Therefore, scraping as a means of land reclamation does not find much favour.

Table 3.1 Management of waterlogged saline soils

Group/sub-group	Technology
Reclamation	Subsurface drainage (Only for waterlogged saline lands)
	On-farm land development including surface drainage
	Salt leaching
Management	
Crop management	Selection of crops and cropping sequences
	Exploitation of varietal differences
	Improved agronomic practices such as higher plant population, going for slightly older seedlings, higher seed rate, addition of FYM, cultivation of green manure crop, if feasible and residue management
Soil management	Land forming/seeding, mulching
Irrigation water management	Shallow depth-high frequency irrigation
	Pre/post-sowing irrigation with fresh water
	Switchover to improved irrigation techniques such as sprinkler and drip irrigation
	Leaching requirement for soils irrigated with saline groundwater
Chemical management	Application of additional nutrients, 25% more nitrogen but all other macro and micro-nutrients on soil test basis
	Application of chemical amendments (only for limited situations)
Rainwater management	*In-situ* rainwater conservation
	Rainwater harvesting and reuse
	Fallowing during summer (monsoon season) to rejuvenate the soils by leaching of salts, especially useful in arid regions

As twin problem of water logging and soil salinity are expanding fast, and emerging realization of the problems in scraping, now-a-days farmers are opting for an altogether different approach. They are transporting normal soil from adjoining areas to raise the field level by about 5-10 cm. Although the cost of such an operation is enormous yet it obviates some of the problems encountered in scraping. While the water table is now deeper by the amount the field level is raised, the disposal related problem is altogether avoided (Fig. 3.1). The long-term sustainability of this practice is questionable as water table is dynamic and in most cases it has a rising trend.

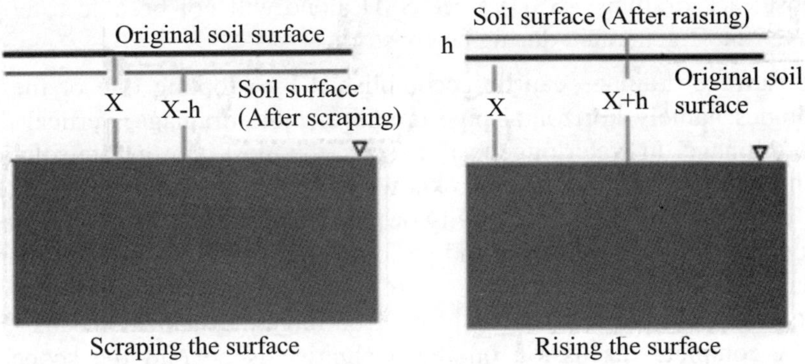

Fig. 3.1 Schematics of scraping and raising the field surface to reclaim saline soils

RECLAMATION COMPONENTS

Drainage

There are two basic types of drainage systems namely surface and subsurface drainage systems. Surface drainage system removes excess water from the surface of land in an orderly manner through improved natural or constructed channels and through shaping of the land surface. The excess water may result from either excess rainfall or excess irrigation. Subsurface drainage, on the other hand, helps to control the water table, which is a pre-requisite to effectively leach the salts from the root zone.

Surface drainage in the monsoon climatic conditions is essentially required for all types of land reclamation programmes. It can be implemented as a part of on-farm land development activities comprising of proper land grading/levelling, construction of dykes (*bunding*) around the farm and field boundaries and preparing a good layout with irrigation and drainage channels of appropriate capacity at the head and at the tail/bottom end, respectively (Gupta, 1992). A grade in the range of 0.05 to 0.2% in the flow direction should suffice to design efficient irrigation and drainage systems. A properly graded field helps in uniform leaching of salts. When surface irrigation methods are used for irrigation, it also facilitates application of shallow depths of irrigation water. Inappropriate land grading on the other hand, will delay the reclamation process or at times even the whole programme might collapse.

When the water table is shallow, soils lack proper aeration because most pore space in the root zone is filled with water. There is excess of carbon-dioxide and lack of oxygen. The other problem arises as a result of inadequate and inefficient leaching because leaching water fails to percolate below the water table. The salts upon leaching continue to remain in the soil profile at a shallow depth. Subsurface drainage systems are, therefore required if the water table is within 1.5-2.0 m below the ground surface and the quality of water is saline in nature. In the monsoon climatic conditions, it is necessary to integrate surface and subsurface drainage (SSD) since SSD alone will not be able to cope with excessive runoff generated during heavy storms.

Subsurface drainage can be accomplished by adopting one of the various technologies namely horizontal pipe drainage, mole drainage, vertical drainage and bio-drainage. In waterlogged saline soils, the most appropriate solution is to go for horizontal pipe drainage also known as simply the pipe drainage (Fig. 3.2 left). It comprises of an outlet (gravity or pumped) for each 40-100 ha area, main drain, collector drain(s), laterals and various other structures notably man holes, junction boxes, drops, water table control structures etc. These systems can be either constructed using semi-mechanical or fully mechanical means (Fig. 3.2 right). A complete discussion on this technique is beyond the scope of this handbook. Briefly, the system details are illustrated in Fig. 3.2. Guidelines on drainage coefficient, drains depth and lateral drain spacing for various soils and agro-climatic conditions of India are presented in Table 3.2 (Gupta, 2011, Gupta, 2019). Readers may also refer any text book on land drainage notably by Gupta (2019).

Table 3.2 Guidelines for SSD design under various agro-climatic conditions in India

Drainage coefficient			Drain depth			Drain spacing	
Climatic conditions	Range (mm d^{-1})	Optimum value (mm d^{-1})	Outlet conditions	Range of drain depth (m)	Optimum depth (m)	Soil texture	Lateral spacing of drains (m)
Arid	1-2	1	Gravity	0.9-1.2	1.1	Light textured	100-150
Semiarid	1-3	2	Pumped	1.2-1.8	1.5	Medium textured	50-100
Sub-humid	2-5	3				Heavy textured/ Vertisols	30*-50

*If the design drain spacing turns out to be less than 30 m, one should go for financial analysis to see if the additional expenditure is recovered in the next 3 years with additional benefits

Besides the relief drainage discussed in foregoing paragraph, drainage may also be required along irrigation network to intercept seepage water so as to prevent development of water logging and soil salinization. For this purpose either the pipe drainage system or for individual farms a non-conventional seepage pond technology developed by CSSRI, RRS, Lucknow can be adopted. On sloping lands, similar kind of system or deep open drains may be required to intercept water before it resurfaces on the ground. All these drainage systems are categorized as interceptor drainage systems.

Fig. 3.2 Schematics of subsurface drainage: plan (left) and a self propelled drain laying machine in operation (right)

Drainage Methods for Coastal Saline Soils

Coastal saline soils are constantly threatened by climatic hazards (cyclones/ floods), shallow saline water table, tidal water inundation, sea water intrusion, poor surface and subsurface drainage, heavy rainfall with deep water logging in *kharif* season and inadequate good quality irrigation water especially in *rabi* season. Although conventional surface and subsurface drainage may be adopted in some cases but more often some unconventional means of drainage are needed. Following 3 methods have withstood the tests and may be employed wherever drainage investigations reveal their potential.

- Tidal bunds with one-way sluice or one way flap gates, which close during high tide but remain open during low tide to drain excess water from the protected areas.
- Farm ponds to harvest, store and use excess rainwater (drainage) during the *kharif* and *rabi* seasons. Besides managing water logging, the technology helps to convert mono-cropped lands to double cropping, minimize soil salinization and significantly higher income to the farmer.
- Land forming and planting methods to cultivate water loving and non-water loving crops or salt tolerant and salt sensitive crops considering soil and water movement in the transformed land configuration.

LEACHING METHODS

Since the initial salinity of saline soils is too high for crop production, the salts are leached down by applying additional quantities of irrigation water. Leaching may occur naturally due to rainfall, a common feature of the monsoon climatic conditions of India (Fig. 3.3). The leaching due to rainfall needs to be significantly enhanced through artificial application of additional water for leaching. The process of application of additional water to leach down the salts is known as leaching. During leaching, salts present in the root zone are dissolved in the water and are pushed beyond the root zone depth. Sometimes and especially in heavy textured saline-alkali soils, flushing method is used to remove the salts. This method is principally different than the leaching process.

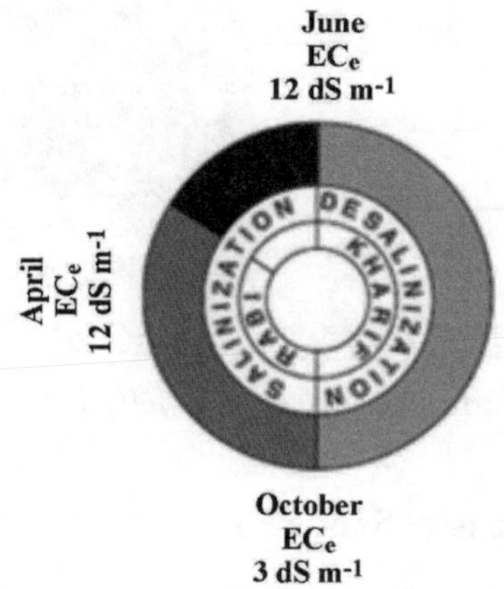

Fig. 3.3 Natural leaching of salts in monsoon climatic conditions of India and soil salinization during winter/summer season

Flushing for Salt Removal

Unlike leaching, which comprises of pushing the salts below the root zone, flushing comprises of surface washing of salts from the top layer of the soil profile. In the latter case, salts move with water as it moves down the sloping land surface. The salt laden water is then drained through a surface drain provided at the tail end. Reeve *et al.* (1955) reported that salts removed through flushing from a 100 m border strip constituted only 1% of the total salt content of the 60 cm deep soil profile. Gupta (1985) observed that except for the first storm or a storm following a long dry spell, no appreciable amount of salt is present in the run-off occurring due to heavy rain in saline soils. El-Gabaly (1971) concluded that flushing may be useful only in areas where there exists a salt crust of highly hygroscopic nature which may prevent use of levelling or other heavy equipment because of low bearing capacity. In such cases, flushing permits better drying of the surface soil. In general, flushing method may be useful only when soil permeability is low or where impermeable layer exists near to the soil surface.

Flushing has been successfully employed in several projects undertaken by *Kharland* Development Board, Gujarat. A combination of flushing and leaching proved beneficial in heavy textured saline alkali soils of the central India. The flushing procedure comprises of the following activities.

- Apply the water and allow it to stand on the land surface for 2-3 days
- Puddle the fields to dissolve soluble salts
- Drain the salty water

Two flushing with an interval of about 7 days proved effective (See chapter 4). A pre-requisite for the efficient functioning of this technique is a well designed surface drainage system. Thus, flushing may not be practicable in land locked fields. High water requirement, inefficient salt removal and several technical constraints limit the use of this technique in land reclamation.

Leaching

The only reasonable way to accomplish salt removal from the root zone seems to be by applying adequate quantities of water on the soil surface, so that the excess salts present in the profile are dissolved and transported below the root zone with percolating water. For efficient leaching, following 3 step approach may be adopted to have uniform and effective leaching.

Field preparation	Level and *bund* the fields
	Carry out shallow or deep tillage. Inversion of the soil may also help in physical reclamation to some extent. In intermittent leaching tillage in between leaching periods helps to maintain infiltration rate
	Make reasonably small size plots by making temporary bunds to ensure uniform leaching
Time of leaching	Select periods when water table is more than 1 m deep
	Select period of low evaporative demand
	Period just before the monsoon season may help to utilize the rainwater for leaching
Depth of reclamation	Select an appropriate cropping pattern as depth of reclamation depends upon the root zone. It is 60 cm for shallow to medium rooted crops or pastures and to more than 2 m for trees. For diversified cropping, the depth of reclamation usually varies from 1 to 1.5 m.

How much to leach?

The salt status of the soil following leaching is normally determined by the tolerance of the crop to be grown. The level of salts should be such that for the planned cropping sequence and water management strategies, salt constituents of concern (salinity, specific ion concentration or alkali) should not exceed the permissible limit. Most comprehensive data on salinity tolerance of crops have been published and form the guidelines for crop selection in reclaimed lands (Maas and Hoffman, 1977; Maas, 1980). Crop tolerance tables included in chapter 5 also provide some information on some selected stations of India. The threshold EC i.e. the EC_e up to which there is no decline in yield, or EC_e at which pre-decided yield reduction is permitted are used to determine the final EC_e of the root zone.

Crop productivity at the end of the season depends upon the initial crop stand. Therefore, most favourable conditions must be created during the germination and initial establishment stages. Since most crops are relatively sensitive to salts at germination or initial establishment stages, the average salinity values reported in crop salt tolerant tables may overestimate the salinity status of the root zone. This point must be kept in view while planning a leaching programme.

Amount of water required for leaching

In an ideal soil system without bypass, without dissolution or precipitation of salts, without salt diffusion or dispersion, the salt concentration of the soil water should drop to the concentration of the applied water when one pore volume (1 PV) of the applied water (1 PV equals one pore space or one pore volume of the soil volume to be leached) has passed through a given depth in the soil profile. Since such ideal situation is never possible under field situations, the actual amount of water required under practical leaching varies from one situation to another. It depends upon many factors, important among them being:

- Amount of salt initially present in the profile
- Desired level of salt in the root zone
- Soil texture
- Type of salts present
- Soil depth to be reclaimed
- Efficiency of the drainage system, if provided
- Method of leaching

As a thumb rule, a unit depth of water passing through the soil profile will remove about 80% of the salts from a unit depth of the soil (Reeve *et al.*, 1955). Field data, however, reveal a wide range depending upon the soil texture and other factors enlisted in this section. The amount of water required to reclaim 1 cm of the soil with continuous leaching may vary from a low value of 0.3 cm to a high value of 2.58 cm. Verma and Gupta (1989) showed that 60 cm of water will suffice to leach 80% of the salts from the top 100 cm of an unstructured clay soil. For leaching of highly saline soil, the amount of water required to pass through the mentioned depth is given in Table 3.3. Leaching curves, discussed in the following section, are commonly used to determine the amount of water required to leach the salts.

Table 3.3 Water required for one time leaching of saline soils

Soil type	Leaching required (cm of water per cm of soil)	Water required to reclaim 60 cm of soil profile (cm)
Coarse texture	0.5-0.6	30-36
Medium texture	0.6-0.8	36-48
Heavy texture	0.8-1.0	48-60

Note: About 80% of the salts present in the soil profile are likely to be leached

LEACHING CURVES

Leaching curve is a useful tool to decide whether desalinization is technically and economically feasible and how much water should be applied to leach the salts to a pre-decided level? Leaching curves represent the relation between the dimensionless depth of water applied (made dimensionless by dividing with depth of the soil) to the dimensionless fraction of the salts leached either on an arithmetic or a semi-log paper. These curves are developed with sufficient accuracy on the basis of field data obtained from simple and rapid leaching tests. Following step wise procedure is used to prepare leaching curves.

- Identify a uniform saline patch in the project area.
- Install double ring infiltrometers at 3-4 places in this uniform patch (Fig. 3.4). Note that set-up is similar to the one used to evaluate infiltration rate (Chapter 8).

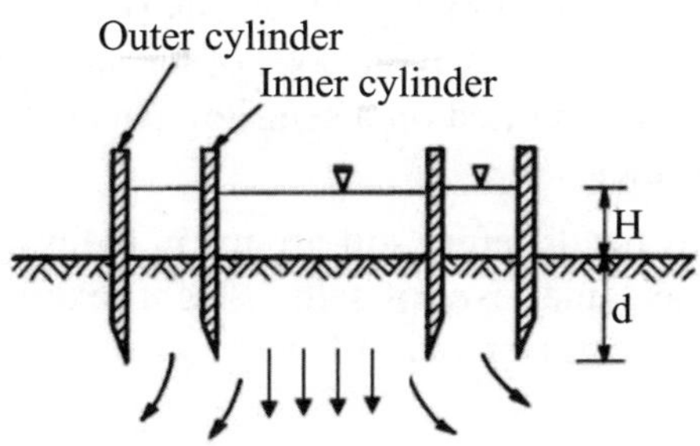

Fig. 3.4 A view of the set-up for preparing leaching curves

- Remove any vegetation etc. from the inner cylinder and take depth wise samples using a tube auger. Fill-up the bore so made so that no preferential flow occurs while conducting the leaching test.
- The water is filled in the two cylinders. The water level in the inner cylinder is kept at a nearly constant level. The water infiltrating in the inner cylinder is measured till the pre- determined amount of water has entered the soil. In each set-up different depths of water are allowed to pass.
- Soil samples in the inner cylinder are drawn after the completion of the test.
- The samples drawn before and after the test are used to evaluate moisture content and salt status in different soil layers. To assess the salt status, bulk soil electrolyte conductivity of different layers is measured.
- An additional sample of 0-5 cm depth is taken in the inner ring of the set-up where maximum amount of water has been applied. The EC_e of this sample is designated as EC_{eq}.

- Now depth of water passing through each layer is calculated by the known amount of water applied, evaporation expected during the period and the water retained in the soil layer. For each set-up one can have 4-6 such layers say 0-15, 0-30, 0-45 and 0-60 cm or 0-10, 0-20, 0-30, 0-40, 0-50 and 0-60 cm. For each layer the depth of water passed is designated as D_w. Division by the depth of the soil for which D_w is calculated, 4-6 values of D_w/D_s are obtained. Similar values are calculated for all the set-ups.
- Similarly, for the same soil layers as used in the previous step, $(EC_e - EC_{eq})/(EC_o - EC_{eq})$ are calculated. Here, EC_o is the initial soil EC_e of the soil layer. Note that if EC_{eq} is neglected, being very small, EC_e/EC_o represents the fraction of salts remaining in the profile (Rf) after the leaching operation. Lower the salt remaining in the profile, higher the leaching.
- The dimensionless plot of D_w/D_s versus $(EC_e - EC_{eq})/(EC_o - EC_{eq})$ is the desired leaching curve. A typical set of leaching curves is shown in Fig. 3.5. If the data are plotted on a semi-log paper, the curves may turn out to be straight lines.
- Similar curves for different soil groups in the project area are prepared. To achieve the same level of salts, a light textured soil can be leached with relatively lesser amount of water than a heavy textured soil (Fig. 3.5).

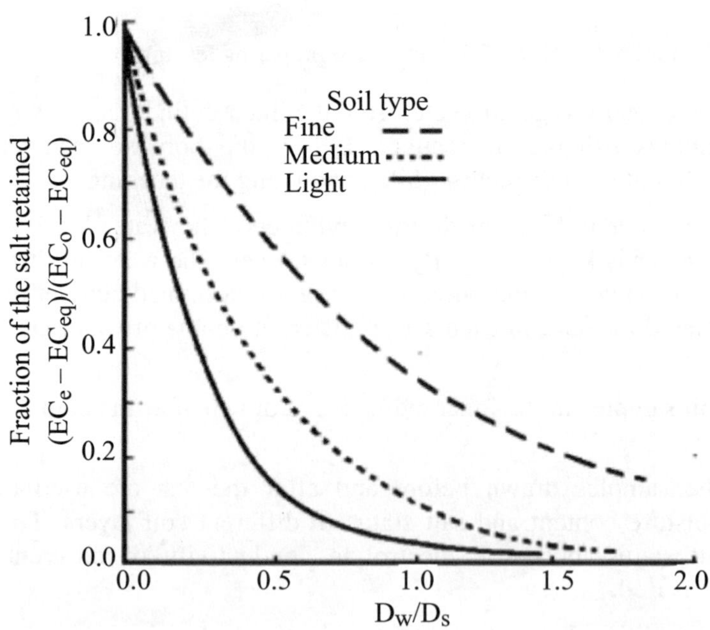

Fig. 3.5 Leaching curves for soils of different textures for the continuous leaching method

To use the leaching curves so prepared, fraction of salts remaining in the profile, Rf is calculated by the following equation:

$$Rf = \text{Crop salt tolerance level/ Initial soil salinity} \quad (3.1)$$

Usually, threshold soil salinity, as previously defined in this chapter is used in eq. (3.1). A higher value than the threshold can be used in eq. (3.1) if some yield can be sacrificed (10% yield reduction is considered normal). It helps to overcome the constraints of water availability, which is a common feature of arid and semi-arid regions. For the fraction so calculated, fraction D_w/D_s is noted and D_w is obtained since D_s, the depth of the soil to be reclaimed is known. To ensure that these test results are applicable to real field condition, one must ensure that

- Same quality of water is used to prepare the leaching curves as is likely to be used during the leaching process.
- Same process of leaching (continuous or intermittent) is adopted that is likely to be used to leach the salts in the reclamation project.

Besides what is discussed, the leaching curves so prepared may not be applicable to drained lands, because of the type of flow encountered in such drained lands. In constructing leaching curves by the method illustrated in the above paragraphs, water flow is vertical. On the other hand, flow in strips near the drains or 1 m below the soil surface in widely spaced laterals with lateral drain depths of 1.5 m or more, is radial. Thus, one has to be quite careful while applying these results to drained lands.

IMPROVING LEACHING EFFICIENCY

A major constraint in saline land reclamation is the limited quantities of water available for leaching. Therefore, several techniques have been developed to improve leaching efficiencies. Some of these techniques are included in this section. But one must be careful in choosing the most appropriate techniques as sometimes no real benefits are obtained with some or all of these techniques.

Continuous and Intermittent Leaching

The most common method of leaching is to store water within the dyked fields. In such a case applied water passes rapidly through the large pores and flushes out the salts through these pores. On the other hand, water is conducted less rapidly through the small pores. Therefore, it takes much longer time to flush out the salts. Clearly, water stored on the surface continues to move through the large pores thereby resulting in inefficient leaching of the soil profile. It has been reported that leaching efficiency can be improved by creating unsaturated conditions through intermittent ponding. This practice minimizes loss of water through large pores and provides greater opportunities for diffusion of salts from immobile or less mobile pockets to the mobile pockets. Thus, it results in thorough leaching of salts from the soil profile. Dahiya *et al.* (1981) reported that nearly 50% saving in water can be made with intermittent leaching compared to continuous leaching in a sandy loam soil. Hoffman (1980) observed that contrary to leaching being a function of soil texture in continuous mode (Fig. 3.5), it is independent of the soil texture in intermittent mode i.e. nearly same amount of water needs to be applied to leach the soils to the same level irrespective of the soil texture. El-Gabaly (1971) reported that although intermittent ponding saved water, it took nearly 40% more time for leaching compared to continuous leaching. Collation of data shows that as much as 30 to 70% water required for leaching in continuous ponding can be saved through intermittent leaching. Later studies have shown that intermittent leaching is not a universal prescription to improve leaching efficiency. Field conditions determine whether intermittent ponding will be beneficial. Nielsen *et al.* (1966) reported that intermittent ponding may not be useful in cases where surface layer is more restrictive than the lower layers. In such cases unsaturated conditions prevail even under continuous ponding. Superiority of intermittent over continuous ponding could not be established in alkali soils (Abrol and Bhumbla, 1973), sandy loam soil (Lefellar and Sharma, 1977) and for reclamation of saline soils formed as a result of irrigation with saline water (Jury *et al.*, 1979). Similar results have been reported for leaching of boron (Beyce, 1972). Surface evaporation of water occurring between leaching may sometime overshadow the increased leaching efficiency brought about by intermittent leaching. Synthesis of data reveals that intermittent leaching may be useful in soils with large bypass (with cracks or structured soils), poor soil permeability, deep groundwater table and during periods of low evaporative demand. On the other hand, even continuous ponding may give reasonably good results in soils with good permeability, unstructured soils, areas having shallow saline groundwater and for periods of high evaporative demand.

Initial Moisture Content and Application Rates

Degree of saturation and the rate of water movement in a given soil also affect the transport of salts through the soil and hence can have profound effect on leaching efficiency (Nielsen and Bigger, 1961, Wilson et al., 1961, Nielsen et al., 1966). All these studies suggested that leaching efficiency can be improved by applying water at slower rates, which can be achieved through intermittent ponding, sprinkling or even under natural rainfall. Later studies however, discounted such inferences and observed that leaching efficiency may not be different when water is applied at a faster or slower rate. It was argued that in the former, the application time is much less than the later. Therefore, in this case leaching status should be measured after allowing redistribution time equal to the time difference between the applications of water under the two set-ups. Following paragraph illustrates this argument.

While studying the effect of initial moisture content, flow velocity and redistribution time on the movement and distribution of salts in soil columns, it emerged that the salt front coincides with the water front in the initially dry soil but lags behind it in the initially moist soil. Immediately after infiltration and after redistribution, the salt peak occurs at a depth above which the total water storage equaled infiltration. But the salt spread in the profile increased as the initial water content increased. It also emerged that immediately following infiltration, the salts are displaced deeper at a rate slower than that of water. Later on, the movement of the salt is faster than that of water. But when the application along with the redistribution time is matched, the salts move deeper with water added at faster than at slower rates. Thokal et al. (1992) observed that more water was required to leach the same amount of salt when water was applied at 1 cm d^{-1} compared to when it was applied at 2 or 3 cm d^{-1}. Dahiya et al. (1984) concluded that there was no significant effect of water application rate and initial soil water content on salt leaching in the experimental soils following infiltration and redistribution.

Conjunctive Use of Saline and Fresh water

If good quality water is either limited or is unavailable at the desired time, conjunctive use of saline and good quality waters for leaching has immense potential especially under the following conditions (Reeve and Doering, 1966).

- The salinity and SAR of the saline water is less than the salinity and SAR of the soil.
- Gradually low EC and SAR water is used for leaching followed by fresh water in the final stage of leaching. Conjunctive use requires that last stage be synchronized with availability of good quality water.

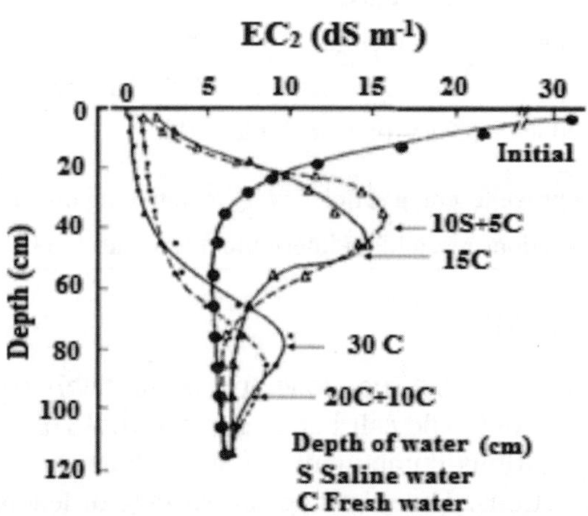

Fig. 3.6 Leaching behavior of saline soils with conjunctive use of saline and fresh water

A number of studies in India have validated the application of this concept and quantified the saving in fresh water with conjunctive use in comparison to when fresh water alone is used (Gupta and Pandey, 1983; Ranade, 1984; Chawla, 1985). Gupta and Pandey (1983) observed that although 18-33% more water (saline + fresh) is needed in conjunctive use, the fresh water requirement depending upon the salinity of the groundwater, initial salinity of the soil and fraction of the salts drained with saline water is reduced by 15–40%. Chawla (1985) also observed that in sandy loam soils fresh water saving in the range of 48.3 to 61.2% can be made although the total water requirement is increased by 37.9 to 63.8%. Typical leaching behavior of a saline soil with conjunctive and fresh water alone is shown in Fig. 3.6.

Additional Practices to Improve Leaching Efficiency

Leaching is the only practical means to reduce soil salinity and bring saline soils under the plough. To reclaim the soils with limited amount of water or to achieve early reclamation, a number of practices can be used. Some useful practices are summarized in Table 3.4.

Table 3.4 Practices to improve leaching of salts

Practice	Effect
Sub-soiling	Improves infiltration rate
	Removes/reduces natural channels
Tillage with inversion	Improves infiltration rate of the soil
	Physically moves the surface layer with high concentration of salts to lower depths
Chemical amendment	Improves infiltration rate
	Controls surface crusting
	Maintains soil structure
Breaking hard pan (Piecing the hard pan with auger)	Improves water movement through the hard pan
	Reduces chances of water table build-up
Irrigation practices	Improve leaching efficiency by maintaining high moisture content
	Situations resembling intermittent leaching can be created

LEACHING REQUIREMENT

The water and salt balances of the root zone are so intricately woven that it is not possible to separate them out. The salt balance of the root zone of irrigated lands can be worked out by various components of water balance multiplied by their respective salt concentrations. The resulting salt storage or leaching from the root zone can be mathematically put into the form of an equation as follows:

$$S = R\,C_r + D_iC_i + G\,C_g + S_1 - D_dC_d - S_2 \tag{3.2}$$

Here salt storage (+) or leaching (-), S, in the root zone is given by the components which add salts to the root zone (+) or those which remove the salts (-). In eq. (3.2) R, D_i, G and D_d are amounts of rainfall, irrigation, groundwater contribution and drainage below the root zone and C is the concentration of water with subscripts describing the respective components. In this equation salts are added to the root zone through rainfall (RC_r), application of irrigation water (D_iC_i), groundwater (GC_g) and direct addition of salts such as through the application of amendments/fertilizer or generated in the root zone through mineral dissolution (S_1). On the minus side are the salts removed by drainage (D_dC_d) and the salt precipitation (S_2). In most cases, the terms S_1 and S_2 are neglected leaving only four terms on the right hand side of eq. (3.2).

Sustainability of irrigated agriculture demands that amount of salts added must not be allowed to accumulate beyond a certain limit. Surplus salts must be removed by the drainage water leaving the root zone. Contribution of rainfall is often neglected because of its low salt concentration. Similarly, if water table is deep, contribution of groundwater can also be neglected. If mass of salts added from weathering of minerals or dissolution of salt deposits, mass of salt added as fertilizers or mass of salts lost as precipitation and mass of salt taken up by crops are neglected, then under steady state conditions with no build up of salinity or leaching of salts, the mass balance of salts in the root zone given by eq. (3.2) is simplified as:

$$D_iC_i = D_d\,C_d \qquad (3.3)$$

From eq. (3.3), it immediately follows that

$$D_d/D_i = C_i/C_d$$

The leaching requirement, LR, defined as the fraction of the irrigation water that must pass through the root zone to maintain soil salinity at some pre-decided level is mathematically written as:

$$LR = D_d/D_i \text{ or } C_i/C_d \qquad (3.4)$$

As electrolyte conductivity (EC) of the solution is a good index of the solution concentration, solution concentration is often replaced by EC such that eq. (3.4) is written as:

$$LR = EC_i/EC_d \qquad (3.5)$$

Here, EC_i and EC_d are the electrolyte conductivities of the irrigation and drainage water respectively. The depth of irrigation water that needs to be applied to ensure both the conjunctive use and LR requirement can be given by the following equation.

$$D_i = D_{cw}\,[EC_d/(EC_d - EC_i)] \qquad (3.6)$$

Here D_{cw} is the consumptive use requirement of the crop.

Although the concept is widely reported in the literature, its limitations must be understood. The steady state conditions assumed in deriving this equation are never met in the monsoon climatic conditions. Under the salinization-desalinization cycles experienced during a year (Fig. 3.2), validity of the LR concept becomes doubtful. In irrigated agriculture, irrigation induced mineral dissolution is significant. Thus, it may not be possible to neglect this component in irrigated lands without causing significant differences in the calculated and actual salt balance. Moreover, where rainfall is significant, it may be useful to use the weighted average salinity of irrigation water and rainfall instead of irrigation water alone in estimating the LR.

In irrigation engineering, we also define D_d/D_i as:

$$D_d/D_i = 1 - E_a \quad (3.7)$$

Such that E_a is the application efficiency. It means LR = 1- E_a. In surface irrigation systems, application efficiencies in India are quite low, the average application efficiency in all irrigation commands of India being as low as 38%. The average water application efficiencies of five irrigation projects in India were observed to be 40% in one, 50% in three and 70% in one (Bos, 1977). On the other hand, leaching requirement of crops varies in a range of 3 to 17%. Hoffman and van Genuchten (1983) reported that LR may be as high as 29% as observed for alfalfa irrigated with saline water of 4 dS m^{-1}. It means in surface irrigation systems full or a part of LR is easily met provided water is applied uniformly over the land surface. Therefore, the concept of LR in the Indian context has limited applications except under presurized irrigation systems. As a testimony, field results of Jain *et al.* (1976) indicate that yield of wheat under irrigation with saline-sodic water (EC = 4.8 dS m^{-1}, RSC = 7.4 meq L^{-1}, SAR = 32) did not increase by application of 25% extra water. Poonia et al. (1974) also observed that levels of leaching (0, 25 and 50%) did not have much effect on wheat and pearl millet grown with saline water.

LAND FORMING: A WAY TO LIVE WITH SALTS

When leaching of salts is difficult either because of limited water or for any other reason, options to live with the salts should be examined while cultivating saline lands. One of the ways to achieve this is by moving the salts away from the seeding/root zone with irrigation and/or by adopting certain land forming systems. The objective is to ensure zones of salt accumulation away from the germinating seeds and plant roots such that seeds and plant roots are exposed to relatively low levels of salts. A combination of irrigation practice and land forming practices can also help to achieve the same objectives.

Furrow irrigation is a free-flow (gravity) surface irrigation system commonly used to cultivate field crops and vegetables. It is used practically on all types of soil, reliefs and slope grades. In a furrow irrigated field, salts accumulate on the ridges while the area at the bottom of the furrow and around it has variable salt status, minimum accumulation being in the zone directly falling under the bottom of the furrow (Fig. 3.7). Thus, seeds can be planted somewhere in between the bottom of the furrow and top of the ridge. Although crops can also be sown in the furrows to take advantage of a wet and less saline zone, yet crops sensitive to water logging may be adversely affected because of periodic water logging stress. Besides, salts may also enter the furrows after a rain shower to adversely affect the crop.

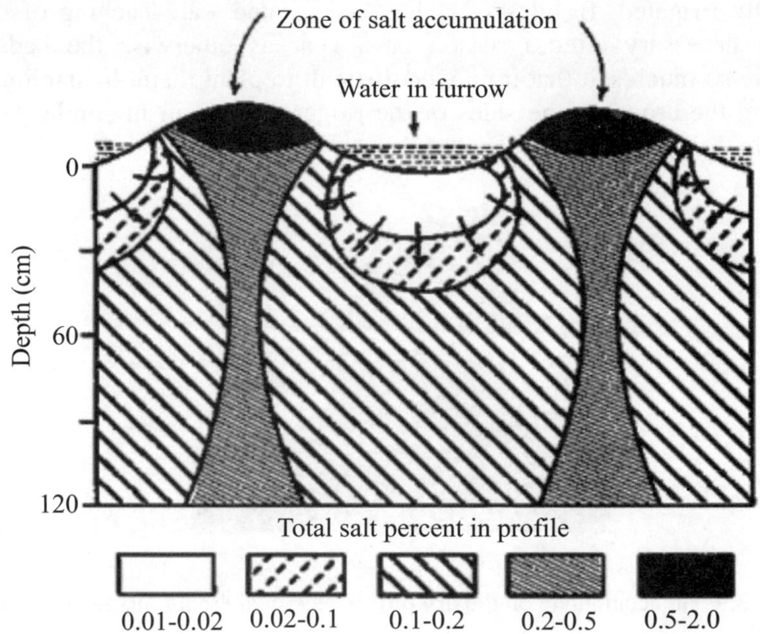

Fig. 3.7 Salt distribution in a furrow irrigation method in saline soils

In a double-row bed system, salts accumulate toward the middle of the bed (Fig. 3.8 left). Thus, plants seeded away from the middle towards the shoulders of the furrow encounter relatively low salt stress. Another modification of this method is also in vogue. Here, a notch at the centre of the ridge is made and planting is done halfway between the wet furrows (Fig. 3.8 right). It may be noted that if water is not applied uniformly to furrows (one furrow receiving more or less water than the other), salts might accumulate closer to one side of the bed to affect the plant growth.

Fig. 3.8 Salt accumulation in double row planting (left) and in furrow top with a notch (right) when each furrow is irrigated

Irrigation methods may also change the salt accumulation sites. In a single-row bed system, salts are pushed across the bed from the irrigated side of the furrow to the dry side in alternate furrow irrigation. Thus, planting towards the shoulder of the wet furrow helps better germination and growth (Fig. 3.9). One must be careful in applying enough water to wet all the way across the bed so that there is no salt build-up in the planted area. Plant injury may still be expected if large amounts of rainfall fill the normally dry furrow and push salts back across the bed. This may also happen if the normally dry furrows are accidentally irrigated. Besides what has been stated, salt leaching of the beds would be necessary after a season or a year as otherwise the beds would accumulate so much salt that it may be difficult to plant them. In slanting ridges, by planting the crops on the sides of the ridges will result in similar beneficial effects (Fig. 3.10).

Fig. 3.9 Salt accumulate on the dry furrow side in alternate furrow irrigation

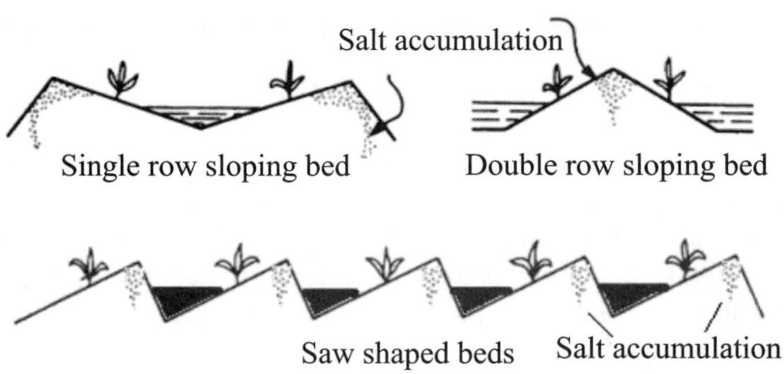

Fig. 3.10 Planting on the sides of slanted ridges to avoid salt stress

Irrigation Methods

One of the reasons of variation in crop yield by different irrigation methods is due to variations in salt and water profiles in the crop root zone. It is because of this reason, high frequency irrigations result in low total stress to the plants. A comparison of furrow-ridge and flat sowing of turnip under surface and sprinkler methods of irrigation for a saline-alkali silty-loam soil resulted in higher concentration of salts up to 30 cm depth in surface irrigation (Yadav and Girdhar, 1977). Turnip yield increased by 25% in sprinkler irrigation, mainly because of lower salinity in the 0-30 cm layer although the soil salinity in lower layers was higher in sprinkler than surface irrigation.

In drip irrigation salts move to the fringes of the wetting front leaving most of the area under the drip salt free (Fig. 3.11). Thus, planting near the emitter helps to avoid salt stress to the germinating seed.

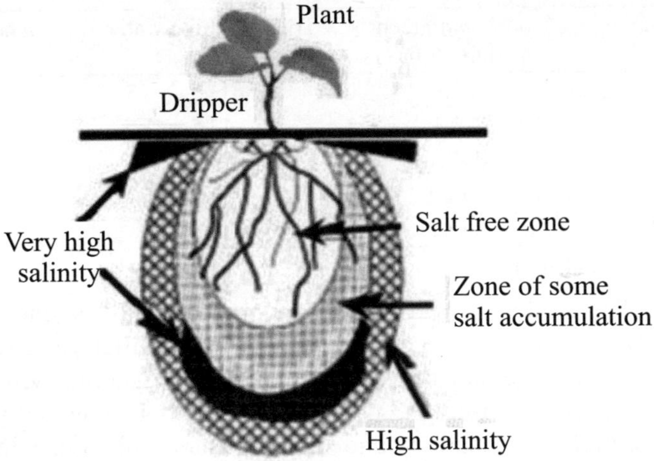

Fig. 3.11 The salt accumulation zone in a drip irrigation system

FIELD APPLICATIONS

The principles and practices of saline land reclamation described in this chapter have been applied to reclaim saline lands in India. Large areas of saline lands having deep water table have been managed using leaching alone. More than 70,000 ha waterlogged saline lands have been reclaimed through subsurface drainage followed by leaching and package of practices included in Table 3.1. This section describes some of the case studies to highlight that saline lands are amenable to reclamation with appropriate application of general principles of land reclamation.

Saline-sodic soils in arid and semi-arid regions having deep water table can be reclaimed through leaching alone. When the reclamation process begins with transplanted rice crop, then pre-transplant leaching may not be necessary. Water applied during field preparation and transplanting is enough to reclaim few centimeters of soil that allows the establishment and survival of young rice seedlings. Fairly good yield can be obtained even in the first year of reclamation (Dahiya *et al.,* 1982). In spite of the fact that these soils have high SAR, there is no need to apply any chemical amendment as SAR of the soil also reduces along with the reduction in salt content. On the other hand, for non-rice crops, pre-plant leaching is essential. A minimum of 30 cm water must be applied to leach down the salts to grow *Sesbania* as a green manure crop (Table 3.5). Additional application of water for leaching was not much beneficial. During the cultivation of *Sesbania*, 33.5 cm of water was used through rainfall and irrigation. The yield of the wheat crop that followed *Sesbania* varied with total amount of water applied for leaching. The yield increase beyond 43.5 cm of water applied was not significant. Apparently, if reclamation begins with wheat, pre-leaching with 40 cm of water may be essentially required to get good yield.

Table 3.5 Effect of application of leaching water on crop yield

Leaching water (cm)	*Sesbania* green manure (t ha^{-1})	Total leaching water (cm)	Wheat (t ha^{-1})
0	33.9	33.5	1.8
10	86.3	43.5	27.4
30	113.1	63.5	33.5
50	119.5	83.5	32.1
70	117.2	103.5	35.0

Source : Dahiya *et al.* (1982)

Gupta (1981) studied the effect of leaching a highly saline soil on pearl millet and wheat crops grown in sequence. The pearl millet yield was very poor when grown without pre-leaching compared to pre-leaching with 20.0 cm of water. The yield improved from 0.3 t ha^{-1} to 1.25 t ha^{-1}. The overall yield was still poor because of high initial salinity and shallow water table. Leaching carried out during the monsoon season helped better growth of wheat following pearl millet crop but only in plots where pre-leaching with 20 cm of water was performed. The major inference of this study was that the reclamation with leaching alone was not sustainable as next *kharif* crop could not be grown. Clearly, for waterlogged saline soils leaching alone is not sufficient for reclamation and needs to be supported by improved drainage to control water table.

Studies conducted at the Sampla farm of Central Soil Salinity Research Institute, Karnal revealed that as soon as a subsurface drainage is installed and other management practices are imposed, leaching and desalinization processes become operative. Salt content of the root zone continues to reduce over the years as long as land reclamation process remains in operation (Gupta, 1985, Rao *et al.,* 1986). Desalinization is quick in semiarid regions where rainfall is on an average 500 mm or more (Table 3.6). In these areas, a salt tolerant crop can be cultivated even in the first year of reclamation. Crops like pearl millet, cotton, mustard, wheat and barley were successfully grown at Sampla (Rao et al., 1986). The yields of these crops continued to increase until yields were comparable to normal soils of the region (Table 3.7).

Table 3.6 Effect of subsurface drainage on soil salinity (dS m^{-1})

Soil depth (cm)	Drain spacing (m)											
	25				50				75			
	A	B	C	D	A	B	C	D	A	B	C	D
0-20	50.7	5.3	2.9	2.7	50.7	8.1	3.6	2.0	46.1	8.3	6.1	4.2
20-40	23.6	4.0	3.3	3.7	19.4	4.7	3.4	2.1	26.4	9.1	7.8	5.0
40-60	19.4	3.7	3.7	1.9	15.8	7.9	4.3	3.1	13.4	9.0	9.4	7.7
60-90	17.0	4.8	5.0	2.0	16.8	11.1	5.8	5.0	11.1	9.4	8.7	7.1
90-120	12.2	7.6	6.2	2.4	15.5	14.3	7.1	4.4	12.6	10.2	10.7	8.4

A=June 1984, B=Oct. 1984, C=Nov. 1984, D=Oct. 1988

Table 3.7 Yield of different crops (t ha^{-1}) as affected by subsurface drainage at Sampla

Crop	1984-85			1987-88		
	25	50	75	25	50	75
Wheat	4.9	4.0	2.5	5.1	5.4	4.9
Barley	4.2	2.9	2.0	2.9	3.1	2.2
Mustard	2.4	2.0	0.9	-	-	-

Similar results were obtained at all the sites where subsurface drainage was used to reclaim waterlogged saline lands. In Karnataka and Andhra Pradesh where rice is cultivated, rice yield increased over the years due to desalinization of the root zone (Fig. 3.12). Large scale projects on land reclamation undertaken in Kota, Rajasthan (RAJAD, 1995), Gohana (Sharma and Gupta, 2006) and Kalayat in Haryana and Sangli in Maharashtra (Dandekar and Chougule, 2010) support these findings.

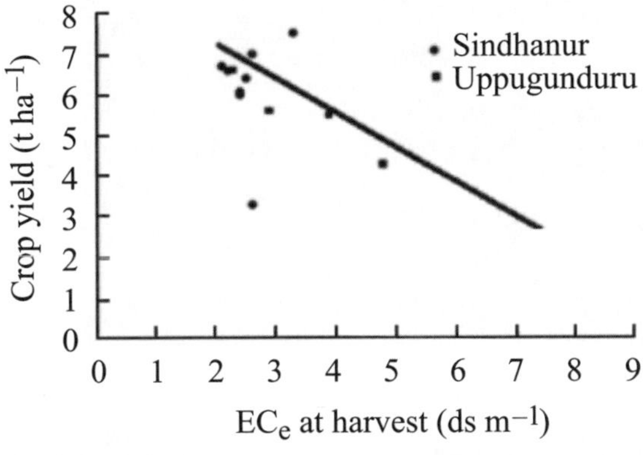

Fig. 3.12 Rice yield as affected by desalinization following land drainage

Seeding on the north-east side of the ridges (Bains and Singh, 1966) proved useful in saline soils, because the salts moved along with water to the drying surface and accumulated there. The EC_e value at south-west, top and the north-east sides of the ridges (60 cm apart, 25 cm high) were 35.4, 24.4 and 6.7 dS m^{-1}, respectively, for the soil initially having EC_e of 15.5 dS m^{-1} (Bains and Singh 1966). Land forming practice to manage a saline alkali soil showed that sowing of sugar beet on one side of the ridge half way between the top and the bottom, opposite to the direction of the sun (northern ridge) gave significantly higher yield than flat and ridge sowing (Dargan and Chhillar, 1973). The salt concentration in the root zone under this set-up of sowing was comparatively less owing to more accumulation at salts on southern face of the ridge due to direct and intense solar radiation on this face. Trench method of planting sugarcane gave significantly higher yield than flat sowing. It is attributed to less concentration of salts, which move to the top of adjoining ridges (Dargan *et al.*,

1973). Sowing on the side of ridges proved superior for maize and wheat in a saline alkali soil at Kanpur, yet the sowing in furrows also gave good results in wheat. Some reduction in maize yield over sowing on the sides of ridges was observed (Table 3.8). It appears that maize being less tolerant to water logging might have suffered due to short-term water stagnation in the furrows.

Table 3.8 Yield of maize and wheat as affected by method of irrigation and seed placement

Treatment	Grain yield (t ha^{-1})	
	Maize (Ganga 5)	Wheat (Sarbati Sonora)
Flat sowing	0.93	2.19
Sowing on the side of ridges	1.85	2.51
Sowing in furrows	1.19	2.52
Sowing in 2 m wide beds	0.95	2.35

■■■■

Chapter - 4

RECLAMATION AND MANAGEMENT OF ALKALI SOILS

Nature and properties of alkali soils are quite different than the normal, saline or acid soils. As a result, these soils pose insurmountable problems during cultivation because of poor physical state and excessive chemical constituents. Unlike saline soils, it is possible to reclaim these soils in the true sense of the word albeit slowly. Reclamation of sub-soils layers usually take several years depending upon the amount of amendment applied, irrigation water used, amount of rainfall and other management options implemented. To enhance the pace of reclamation and for effective reclamation, management of these soils is quite important during the first few years of reclamation. The reclamation and management components of these soils are listed in Table 4.1.

Table 4.1 Reclamation and management package for alkali soils

Group/sub-group	Technology
Reclamation	On-farm land development including surface and vertical drainage
	Application of chemical amendment
	Flushing#/Leaching
Management	
Crop management	Selection of crops and cropping sequences
	Exploitation of varietal differences
	Improved agronomic practices such as higher plant population, older age of seedling, higher seed rate, skipping puddling etc. including cultivation of green manure crop after the wheat crop
Soil management	Land forming/seeding
Irrigation water management	Shallow depth-high frequency irrigation regime
	Switchover to improved irrigation techniques
Chemical management	Application of additional nutrients/green manure/FYM
Rainwater management	*In-situ* rainwater conservation
	Rainwater harvesting and reuse (helps in drainage)

#Flushing is adopted only in heavy textured soils with accumulation of excessive salts on the soil surface

CHEMICAL AMENDMENTS

Simply put, a chemical amendment is a material that is added to the soil to change and improve the soil's physical or chemical properties that aid in plant growth. Amendments, including those that improve soil physical properties, work through chemical, biological, and physical modes of action. Even fertilizers are materials that contain useful amounts of one or more plant nutrients and helps in better plant growth but in a practical sense, these are not categorized as amendments. For the purpose of this chapter, a chemical amendment is one that is used to reclaim alkali soils to overcome the adverse effects of high sodium and/or high pH. The chemical amendments neutralize the pH of the soil, react on the free sodium carbonate, and replace the exchangeable sodium by calcium. The initial soil ESP, time span within which one wants to reclaim the land and presence of *kankar* (impure $CaCO_3$) in the soil help to decide the kind of amendment to be used. These factors together determine the effectiveness of a chemical amendment. Besides, the success of a chemical reclamation programme depends to a great extent on the quality of irrigation water and the depth to water table. The good quality irrigation water and deeper depth to water table helps to achieve complete, early and sustainable reclamation. This chapter discusses various kinds of amendments, their availability, methods of application and their response in land reclamation.

KIND OF AMENDMENTS

The chemical amendments are broadly classified in three categories, namely

Soluble calcium salts, viz. calcium chloride and gypsum which directly supply soluble Ca for the replacement of exchangeable Na.

Acidifying materials, viz. sulphuric acid and elemental sulphur which due to their acidulating action and/or increasing the partial pressure of CO_2 supply Ca indirectly by dissolving calcite natively found in alkali soils. The Ca so mobilized is used to replace Na from the exchange complex. Other amendments under this category besides sulphur and sulphuric acid are: iron sulphate, aluminium sulphate and lime-sulphur

Organic materials viz. rice husk, FYM and green manure etc. Compost and other organic materials improve soil physical structure by two distinct mechanisms – by stimulating microbial activity, which helps in soil aggregation, and at high application rates, by directly decreasing soil density. These also help to improve soil structure as well as dissolve the native $CaCO_3$ albeit weakly. The reclamation efficiency is usually low with these kinds of amendments. Organic materials together with inorganic amendments are more cost effective, hasten the reclamation process and increase the yield. Therefore, integrated use of inorganic and organic amendments is desirable.

Industrial wastes are also used as amendments. Depending upon their constituents, these amendments work on similar principles as the above mentioned three kinds of amendments. Amongst the commonly used materials under this category are: pyrites, distillery spent wash, dairy waste, phosphogypsum and press mud. These amendments may also contain some impurities/ contaminants and therefore, the adverse effect of these contaminants on soil or crop quality should be investigated.

CHEMICAL REACTIONS OF AMENDMENTS

Whenever a chemical amendment is applied to an alkali soil, depending upon its constituents and constituents present in the soil, chemical reaction occur that helps to replace the exchangeable sodium from the exchange complex. Some of the chemical amendments and their chemical reactions are illustrated as follows.

Gypsum

$$\underset{\text{Alkali soil}}{2\,Na-clay + 2CaSO_4} \rightarrow \underset{\text{Normal soil}}{Ca-clay + Na_2SO_4\downarrow} \quad (4.1)$$

Phosphogypsum

In addition to the mineral gypsum, large quantities of byproduct gypsum are available from different industries. The phosphogypsum is produced when rock phosphate is treated with sulphuric acid as per the following reaction:

$$\underset{\text{Apatite(rock phosphate)}}{Ca_{10}(PO_4)_6F_2} + 10H_2SO_4 + 20H_2O \rightarrow \underset{\text{Phosphogypsum}}{10\,CaSO_4\cdot 2H_2O} + 6H_3PO_4 + 2HF \quad (4.2)$$

$CaSO_4\cdot 2H_2O$ reacts with the soil in the same manner as gypsum (eq. 4.1).

Pyrites

The low quality pyrites having different percentage of sulphur (15-30%) classified as agricultural grade is normally used in the reclamation of alkali soils.

$$2FeS_2 + 2H_2O + 7O_2 \rightarrow 2FeSO_4 + 2H_2SO_4 \quad (4.3)$$

$$CaCO_3 + 2H_2SO_4 \rightarrow CaSO_4 + CO_2 + H_2O \quad (4.4)$$

$$2\,Na–clay + CaSO_4 \rightarrow Ca–clay + Na_2SO_4\downarrow \quad (4.5)$$

If the soil is non-calcareous, the reaction will proceed as:

$$2\,Na–clay + H_2SO_4 \rightarrow 2H–clay + Na_2SO_4\downarrow \quad (4.6)$$

$$2\,H–clay + CaCO_3 \rightarrow Ca\text{-}clay + CO_2 + H_2O \quad (4.7)$$

It may be noted that sulphur present in the pyrites is in a highly reduced state. It readily reacts with the oxygen of the air and forms sulphur compounds of a higher oxidation state. Both the products formed directly or indirectly react with soil. As against this, the elemental sulphur has to be oxidized to sulphuric acid by the sulphur oxidizing bacteria like *Thiobacillus thiooxidans,* etc., and its oxidation therefore, is quite slow in alkali soils.

Sulphur and Sulphuric Acid

Mineral sulphur is converted to H_2SO_4 through the process of biological oxidation by *Thiobacillus thiooxidens* as per the following reaction:

$$2\,S + 3\,O_2 \rightarrow 2\,SO_3 \text{ (microbiological oxidation)} \quad (4.8)$$

$$SO_3 + H_2O \rightarrow H_2SO_4 \quad (4.9)$$

At high pH commonly encountered in alkali soils, oxidation in this manner is a very slow process. As such, the relative efficiency of mineral S as an amendment is low. Further reaction will be as per eq. (4.4) and eq. (4.5) in the case of pyrites in this section. If the soil is non-calcareous, the reaction will follow eq. (4.6) and (4.7).

The H_2SO_4 and $CaCO_3$ react in the following manner.

$$H_2SO_4 + 2CaCO_3 \rightarrow CaSO_4 + Ca(HCO_3)_2 \quad (4.10)$$

Under these conditions both the $Ca(HCO_3)_2$ and $CaSO_4$ react with exchangeable sodium.

Following chemical reactions occur when H_2SO_4 is applied as an amendment to an alkali soil.

$$H_2SO_4 + CaCO_3 \rightarrow CaSO_4 + CO_2 + H_2O \quad (4.11)$$

$$2\,Na - clay + 2CaSO_4 \rightarrow Ca - clay + Na_2SO_4 \quad (4.12)$$

If the soil is non-calcareous, the reaction will follow eq. (4.6) and (4.7).

Ferrous Sulphate

$$FeSO_4 + H_2O \rightarrow H_2SO_4 + FeO \quad (4.13)$$

In a calcareous soil further reactions will occur as per the eqs. (4.4) and (4.5) and for non-calcareous soils as per eq. (4.6) and eq. (4.7).

Lime sulphur

$$CaS_5 + 8\,O_2 + 4\,H_2O \rightarrow CaSO_4 + 4\,H_2SO_4 \quad (4.14)$$

Both $CaSO_4$ and H_2SO_4 will react in alkali soils as per eq. (4.1) and eq. (4.11).

For a non-calcareous soil, the reaction may proceed as:

$$10\,Na - clay + 4H_2SO_4 + CaSO_4 \rightarrow 8H - clay + Ca\text{-}clay + 5Na_2SO_4 \quad (4.15)$$

$CaCO_3$ as an Amendment

As a general rule, almost all alkali soils contain $CaCO_3$ in sufficient amount in the root zone depth commonly referred as *Kankar* ($CaCO_3$ concretions). Therefore, any additional application of $CaCO_3$ may not be useful. Moreover, ground limestone is an effective amendment only in soils having pH below about

7.0 because its solubility rapidly decreases with increasing soil pH (Table 4.2). Therefore, there is no point in adding this as an amendment for alkali land reclamation. On the other hand, efforts could be made to increase the solubility of slowly soluble $CaCO_3$ concretions (13.1 mg L^{-1}). It can be achieved through increase in the concentration of CO_2 so that $CaCO_3$ is converted into calcium bicarbonate whose solubility is much higher, being in the range of 60 to 140 mg L^{-1}. Similarly, lowering the soil pH also increases the solubility of $CaCO_3$ (Table 4.2). Thus, if by any means, CO_2 concentration in the soil can be increased or the pH of the soil can be decreased, the low and medium degree alkali soils can be reclaimed naturally. The most practical means of reducing pH and permanently increasing the CO_2 concentration in the soil is to intensify the microbiological processes through the decomposition of organic substances ploughed into the soil in the form of green or farm yard manure. The application of small amounts of gypsum or other acidifying materials and the conservation of rainwater during the monsoon season can also help to speed up the process of natural reclamation. Gupta and Abichandani (1970) reported that in a highly calcareous sodic soil, the ESP was reduced from 59.5 to 18.0 and from 34.9 to 11.0 due to leaching alone caused by the rainfall (350 to 450 mm) in Western Rajasthan. In addition to the remarkable changes in ESP, pH and the total concentration and composition of the soluble salts also changed favourably. Besides, soils that have high exchangeable sodium and also contain appreciable exchangeable hydrogen resulting in having an acid reaction are amenable to $CaCO_3$. In India, only limited soils of this character have been reported in the Sunderbans. The reaction of $CaCO_3$ will be as follows.

$$2\ \text{H–clay} + CaCO_3 \rightarrow \text{Ca-clay} + CO_2 + H_2O \tag{4.16}$$

Table 4.2 Solubility of $CaCO_3$ (mg L^{-1}) as affected by the CO_2 content and pH of the soil

CO_2 content in soil air (% by volume)	pH	Solubility of $CaCO_3$
0.00	10.23	13.1
0.03	8.48	63.4
0.30	7.81	133.4
1.00	7.47	202.9
10.00	6.80	470.0
100.00	6.13	1098.6

Source: Kovda et al. (1973)

Equivalence of Amendments

The equivalent quantities of the chemical amendments other than gypsum can be calculated from equivalent weights of the amendments (Annexure I). Equivalent amendment to 1 unit of gypsum is obtained by multiplying the gypsum quantity with the ratio of the equivalent weight of the amendment to the equivalent weight of gypsum. For example equivalent weight of sulphuric acid and gypsum are 49 and 86 respectively. Therefore, 1 unit of gypsum is equal to 0.569 units of H_2SO_4 (49/86 = 0.569). Similar procedure is used to calculate the equivalence to sulphur. Some of the equivalent values for other amendments to gypsum and sulphur are given in Table 4.3.

Table 4.3 Commonly used amendments and their equivalent amendment values

Amendments (Chemical formula)	Tons of amendment material equivalent to	
	1 ton of pure gypsum	1 ton of pure sulphur
Gypsum ($CaSO_4.2\ H_2O$)	1.00	5.38
Calcium chloride ($CaCl_2.2\ H_2O$)	0.86	--
Calcium nitrate ($Ca\ (NO_3)_2.2\ H_2O$)	1.06	--
Press mud (lime–sulphur, 9% Ca+24% S)	0.78	4.17
Sulphuric acid (H_2SO_4)	0.57	3.20
Iron sulphate ($FeSO_4.7\ H_2O$)	1.62	
Ferric sulphate ($Fe_2(SO_4)_3$)	1.09	5.85
Aluminium sulphate ($Al_2\ (SO_4)_3.18\ H_2O$)	1.29	6.34
Sulphur (S) **	0.19	1.00
Pyrites ((FeS_2, 30% S) **	0.63	

*The above quantities are based on 100% pure materials. If the purity is less than 100%, then necessary corrections must be made. Thus, if gypsum is only 80% pure then the equivalent quantity to be applied would be 1.25 times of quantities mentioned in the Table.

** For sulphur and pyrites, 100% oxidation is assumed.

Gypsum Requirement (GR)

Gypsum application in India has been known from the year around 1898, yet its practical application in land reclamation never picked up until establishment of CSSRI. The major thrust of research at the institute has been to bring economy in the application of amendments. The first step to reduce the cost was to establish and fine-tune the GR methodology. A number of methods are available to calculate the gypsum requirement of the soils (See Chapter 7). The list of the methods is given as under.

- Use of theoretical relation
- Schoonover (1952)
- Modified Schoonover method (Abrol *et al.*, 1975)
- pH versus GR graph

Schoonover (1952) method is most commonly applied to determine the GR. Nonetheless, studies at CSSRI have resulted in several improvements in this method.

- It has been proved that for a shallow rooted crop such as rice, it may be prudent to only reclaim the top 0-15 cm profile rather than the 30 cm recommended in the original Schoonover method. It essentially helps to nearly halve the GR.
- Field studies have further shown that only 50% of the GR determined for the 0-15 cm soil profile is sufficient to reclaim alkali soils in India. From economic angle these observations proved to be major milestones in the proliferation of this technology.
- Alkali soil should be leached before drawing soil samples to assess the GR. This suggestion has emerged from several studies wherein it has emerged that GR determined with the soil samples without leaching are higher than soil samples drawn after leaching (Table 4.4, Arora et al., 2015). A part of the gypsum applied is consumed in neutralizing the free Na_2CO_3. It is therefore, recommended that the soil samples drawn after leaching may be analyzed for GR as per Schoonover method. Otherwise Abrol *et al.* (1975) method should be used, in case soil samples are drawn without leaching. As a corollary, the soils should also be leached before gypsum is applied to the soils. A higher proportion of the calcium supplied by the amendments is then used in exchange of Na from the exchange complex.

Table 4.4 Effect of flushing on gypsum requirement

Location	Soil pH_2	Soil EC_2 (dS m^{-1})	Gypsum requirement (meq $100g^{-1}$)	
			Without leaching	After leaching
Koithar	10.1	8.18	5.70	1.40
Kausambhi	9.2	0.74	3.70	2.90
Rasoolpur	8.9	2.43	1.60	1.10
Lalganj	10.0	1.28	5.80	3.40
Sandila	10.3	2.50	5.30	3.70
Dhaura	10.1	4.04	4.50	2.20
Vishnupur	9.8	2.25	5.80	4.80

Quantity of Gypsum Required for Reclamation

The quantity of gypsum obviously depends upon the initial ESP of the soil and the level to which it has to be reduced, presuming that the irrigation water is of good quality. The level to which the exchangeable sodium content of the soil is to be reduced depends upon the soil texture and the first crop to be grown after amendment application. For example, if the soil is initially fully saturated with the exchangeable sodium (ES = 10 to 15 meq 100 g^{-1}, ESP = 95 to 100), it may suffice to reduce the ESP to about 50 if tolerant crop like rice (transplanted) is to be grown. Likewise, ESP has to be reduced to 30 if semi-tolerant crops like wheat, or to 10 if sensitive crops like gram, are to be cultivated. Abrol *et al.*

(1973) and Bhumbla and Abrol (1975) exploited the synergies of such options and observed that satisfactory yield of rice, wheat and barley can be obtained by adding gypsum @ 50% of gypsum requirement as determined in the laboratory by Schoonover's method for the 0-15 cm depth.

Fineness

Gypsum is quite cheap in India; it's cost at the mining site being negligible. The degree of fineness to which gypsum is grounded adds to its cost. According to the ISI specification, the whole of the material must pass through 10 mesh and at least 50% of it must pass through 60 mesh (0.25 mm) sieve. This standard pertains to fineness of about 30 mesh. Grinding to a fineness of 30 mesh contributes about 15% of the total cost of gypsum. The cost can be reduced if we adopt a standard where gypsum of 10 mesh fineness is used in reclamation programmes as suggested in several studies (Kovda *et al.*, 1973). The studies at CSSRI suggest that gypsum passed through a 2-mm sieve, having a wide particle size distribution, is more effective to reclaim alkali soils having appreciable quantities of Na_2CO_3 in the soil solution (Chawla and Abrol, 1982).

Solubility

The rate of reaction of gypsum is limited by its moderate solubility, its solubility in water being about 0.25% at ordinary temperatures. The presence of sodium and chloride ions in the water helps to increase the solubility of gypsum, whereas calcium and sulphate ions tend to decrease its solubility. It has been conventionally recognized that an application of 90 to 120 cm of irrigation water is required to dissolve 10 to 12 t ha^{-1} of high-grade gypsum (Richards, 1954). Recent studies have shown that it is true only when gypsum is applied to normal soils. Against this, the solubility of gypsum in alkali soil is higher depending upon pH, and ESP of the soil. It is because the soil exchange complex acts as a sink for the Ca released in the soil. It has been observed that in alkali soils, the solubility of gypsum is ten times more (Oster and Halvorson, 1978). Abrol *et al.* (1979) reported that solubility of gypsum increases with increase in ESP of the soil as per the following reaction.

$$Y = 0.0186\,X + 0.18 \quad (r = 0.98) \tag{4.17}$$

Here, Y is the mean gypsum solubility (%) and X is the ESP of the soil.

Hira *et al.* (1981) reported that gypsum solubility increases with increasing gypsum requirement of the soil.

$$Y = 0.46\,N + 2.1 \tag{4.18}$$

Here, Y is the mean gypsum solubility (%) and N is the gypsum requirement of the soil.

On this basis, Hira and Singh (1980) observed that the amount of water required to dissolve the added gypsum to reclaim alkali soils does not increase with the amount of gypsum added provided it equals the gypsum requirement.

They also observed that only 4 cm of water is sufficient to dissolve gypsum (<0.26 mm size) but may require about 26 cm of water to leach down the exchanged products.

Frequency of Application

Single application of gypsum @ 50% of GR has been found to be more efficacious than repeat applications over a 2-3 years period. Even single application of 25% GR (6.5 t ha^{-1}), before the first crop of rice, has proved as effective as the higher and repeated doses in subsequent years for a highly alkali soil. The pH_2 came down from 10.7 in the first year before the rice crop to 9.5, 8.6 and 8.4 after the wheat crop in the first, second and third year, respectively. It shows therefore, that if reclamation has to proceed with the rice as the first crop, application of gypsum even @ 25% GR would be enough, if average yields for 3 or more years are taken as the basis of recommendation. However, the yields of rice and wheat are low in the first year by applying gypsum at the rate of 25% GR as compared to 50% GR.

Self Reclamation

With continuous use of high quality irrigation water following good irrigation and cropping practices, reclamation in the sub-soil usually proceeds with time because of the following processes that come into operation:

- Almost all alkali soils contain sufficient reserves of *Kankar,* whose solubility increases over the years through various processes. Thus, self reclamation process begins as per the following reactions.

$$\text{Na} - \text{clay} + \text{H}_2\text{O} \rightarrow \text{H} - \text{clay} + \text{NaOH} \quad (4.19)$$

$$2\text{H} - \text{clay} + 2\,\text{NaOH} + \text{CaCO}_3 \rightarrow \text{Ca} - \text{clay} + \text{Na}_2\text{CO}_3 + 2\,\text{H}_2\text{O} \quad (4.20)$$

- The irrigation water of high quality, usually available from the canals, contains Ca eight to ten times higher than Na. This results in constant replacement of Na with Ca on the exchange complex particularly after the application of gypsum
- The use of nitrogenous fertilizers, particularly $(NH_4)_2SO_4$ is very effective in lowering the soil pH. This also exerts a marked influence in releasing more calcium from $CaCO_3$.
- Consequent upon the start of the process of reclamation through cultivation of crops, the plant roots evolve CO_2 which helps in the conversion of $CaCO_3$ into $Ca(HCO_3)_2$. The solubility of the later being more than the former
- Because rainwater is acidic in nature due to its CO_2 content, the natural reclamation of the soil is aided through conservation of rainwater.

In view of the foregoing analysis, it can be generalized that in a particular situation small amounts of gypsum (25% GR) can act as a trigger to the process of reclamation of alkali soils provided

- Good quality water is available for irrigation
- The rains are appreciable and effective
- The soils contain sufficient reserves of $CaCO_3$
- Adequate amount of nitrogenous fertilizers are used
- Rice (transplanted) is grown as the first crop.

The process of natural reclamation however, is subject to the land being cropped continuously. The land may not be left fallow for prolonged periods.

Rice Crop for Reclamation

A major contribution of the CSSRI to the land reclamation programme has been to begin the reclamation process with rice as the first crop. Rice crop helped the reclamation process in the following ways.

- Rice crop is able to tolerate standing water and the water used by the crop is simultaneously used to dissolve and leach excess soluble salts and exchanged products.
- Crop is quite tolerant to high ESP. Besides the pH of the soil is reduced because of the standing water in the fields.
- Chhabra and Abrol (1977) from laboratory studies observed higher concentration of HCO_3 in the leachate in the presence of rice crop than without it. They postulated that it confirms the reaction of

$$CaCO_3 + CO_2 + H_2O \rightarrow Ca(HCO_3)_2 \tag{4.21}$$

$$2Na - clay + Ca(HCO_3)_2 \rightarrow Ca - clay + 2\,NaHCO_3 \tag{4.22}$$

Leaching

Except where pyrites and sulphur is used, alkali soils should be leached immediately following amendment application. Leaching water, besides dissolving and carrying the amendment downward, also removes the exchanged products from the root zone. Leaching for 7-10 days before transplanting the rice crop is recommended because additional leaching is accomplished by the water standing in the rice crop.

Questions on intermittent and continuous leaching to reclaim alkali soils have been addressed in several studies. It has been observed that intermittent leaching has no particular advantage in alkali soils. It is argued that in alkali soils where ESP of the surface soils limits the infiltration rate, unsaturated flow conditions prevail in the profile. Therefore, no additional benefits accrue due to intermittent leaching. Moreover, in the first year of reclamation, time window is small and therefore, continuous leaching is recommended for leaching of alkali soils if gypsum is used as an amendment.

Although sulphur has not been used so far on a large scale in India, experience with pyrites reveal that sufficient time should be allowed for oxidation, before leaching is initiated. Application of amendment on a moist soil, 4-7 days period of oxidation followed by leaching preferably intermittent has been found useful for pyrite.

Even organic amendments require standing water on the fields following their application. Organic matter decomposed under aerobic conditions has been found to be inferior to their decomposition under saturated (anaerobic) conditions. Singh (1974) observed that efficiencies of barley straw and *Sesbania* in terms of reduction in ESP and improvement in soil physical properties was less under aerobic than anaerobic conditions. Thus, to get maximum benefits from organic amendments, it is advisable to maintain continuous ponding during the course of decomposition of the organic materials added to an alkali soil.

Speed of Reaction of Amendments

The choice of a chemical amendment in the Indian context mainly depends upon its cost. Besides time required for its reaction in the soil and its logistics of transport and application also determine the potential of an amendment. For example, calcium chloride may be the most readily available source of soluble calcium, but is seldom used due to its high cost. On equivalent basis, sulphuric acid proved to be better than gypsum for barley crop but its application poses several limitations (Yadav, 1973). Sulphur is a slow acting amendment as discussed previously in this chapter. Considering all these factors, gypsum either as mineral gypsum or phosphogypsum are commonly used as chemical amendments in India. Although, we have seen that solubility of gypsum in alkali soils is much higher yet the application of additional water during first few years either by selecting water loving crops or cultivating green manure crop helps to increase the pace of reclamation. For example, growing of *Dhaincha* (a green manure crop following rice wheat rotation in the first year of reclamation) helps in speedy reclamation because of additional water applied during its cultivation and the addition of organic matter to the soil. Besides, cultivation of fodder crop *Shaftal* (*Berseem* is less tolerant to high ESP and therefore can be taken up from second year onward) enhanced the pace of reclamation because of more water required to grow this crop and being leguminous in nature. Depending upon climatic conditions irrigations to *Shaftal* are applied at intervals of 10-15 days in winter season as compared to wheat where irrigation interval is 21 days or more. Application of gypsum in the nano-form is being researched to improve the pace of reclamation or as a cost cutting exercise (Santhosh and Thiyageshwari, 2018). In general, it takes about 3 years before optimum crops of rice and wheat are obtained in gypsum treated fields although it takes 8-10 years before one can cultivate crops of one's choice including sensitive crops.

Most contrasting results emerged in respect of pyrites as a chemical amendment. While some reports described it an equally good amendment as the mineral gypsum but others rated it much inferior to it. A detailed investigation revealed that the efficiency of pyrites is dependent on its water-soluble sulphur content at the time of application. While pyrites with 5.5% and 8% soluble sulphur is as effective as gypsum, freshly mined pyrite (water-soluble S 1%) is quite inefficient in reclaiming alkali soils (Table 4.5). To increase the water-soluble sulphur content of pyrite, freshly mined pyrite needs to be stored under moist conditions by sprinkling water over the bags under a rain shelter. With the enrichment of indigenous iron- and sulphur-oxidizing bacteria of pyrite, the water-soluble sulphur increases to 5% within a period of 6 months. No such increase has been observed when pyrite is stored under dry conditions. A simple field test can be performed to test the pyrite quality. Dissolve one-gram pyrite in 100 ml distilled water and check the pH using a pH meter or a litmus paper. If the pH is equal or less than 2.5, it may be applied otherwise it should be stored for some more time (Sharma and Swarup, 1990).

Table 4.5 Effect of gypsum and pyrite on yield and soil properties

Treatment	pH (1:2) (0-15 cm)		ESP (0-15 cm)	Yield (t ha^{-1})	
	After paddy	After wheat		Paddy	Wheat
Control	10.0	10.1	69.1	3.3	0.0
Gypsum (50 % GR)	9.5	9.3	50.2	5.2	3.3
Gypsum (75 % GR)	9.3	9.3	32.9	5.3	4.2
Pyrite I* (50 % GR)	9.9	9.9	60.2	4.6	0.6
Pyrite I (75 % GR)	9.7	9.8	55.8	4.4	0.4
Pyrite II* (50 % GR)	9.7	9.5	48.5	5.1	2.9
Pyrite II (75 % GR)	9.2	9.2	30.9	5.3	4.1
Pyrite III* (50 % GR)	9.7	9.7	50.0	5.6	3.4
Pyrite III (75 % GR)	9.2	9.3	29.7	5.7	3.9

Pyrite I, II and III contained 2.8, 5.5 and 8.0 soluble sulphur at the time of use

FIELD TESTS

The principles and practices of alkali land reclamation listed in Table 4.1and described in various sections of this chapter have been applied to reclaim about 2.07 M ha area so far. It has given additional food grains production of about 16 million tons. Although, it is not the objective of this handbook to describe the field or laboratory experimental results, but just to illustrate some of the statements made in this chapter few data sets are included for discussion. Following points emerging from various experimental results are worth mentioning.

- Extensive work done has shown that 10 to 15 t ha^{-1} of mineral gypsum containing 70% $CaSO_4.2H_2O$, which approximates about half of the actual GR determined by Schoonover's method (0-15 cm depth), is sufficient to reclaim the surface 15 cm soil provided one begins with transplanted rice as the first crop (Table 4.6). It is keeping in view the shallow root system of the crop and the high tolerance of transplanted rice to high soil ESP. Besides, water stored in the rice fields helps to reduce the soil pH. Moreover, its roots capacity to facilitate the soil water to dissolve soil $CaCO_3$ also helps to reduce the soil exchangeable sodium.

- Rice yield within 3 years even in control plot without any amendment are reasonably high although differences persist in the wheat yield. At other places, where soils are saline-alkali in nature and are of relatively finer texture, yield differences remain even in the case of rice yield. Over time, ESP of the surface as well as deeper layers gets reduced to an extent that other crops can be cultivated after 3 years of rice-wheat cropping.
- Organic amendments act slowly but in about 2-3 years may reclaim the lands. Organic amendments along with chemical amendment @ 25% of GR act synergistically and proved to be equally good as the chemical amendment @ 50% of GR alone.

Table 4.6 Yield (t ha^{-1}) of crops as affected by gypsum treatment of an alkali soil

Gypsum (t ha^{-1})	*Sesbania** 1970	Wheat 1970-71	Rice 1971	Wheat 1971-72	Rice 1972	Wheat 1972-73
0.0	0.44	0.0	4.39	1.52	7.18	1.35
7.5	3.99	1.89	6.21	2.79	7.23	1.96
15.0	4.35	3.49	6.39	3.45	7.13	2.43
22.5	4.70	4.16	7.08	3.72	7.17	2.26
30.0	4.50	3.79	6.66	3.98	7.03	2.74
LSD at P = 0.05	0.51	0.64	0.81	0.70	NS	0.50
Normal yield	-	4.5	7.0	4.5	7.0	4.5

* Green matter yield; Initial soil pH = 10.3; ESP = 94 in surface 15 cm soil

Source: Abrol and Bhumbla (1979)

A comparative evaluation of mineral gypsum and phosphogypsum (PG) revealed that at equivalent rates ESP values were lower in phosphogypsum treated fields as compared to mineral gypsum treated ones. Differences persisted even after 3 years after reclamation (Nayak *et al.*, 2011). Higher solubility of phosphogypsum along with increased availability of P over mined gypsum were attributed to lower ESP levels (Table 4.7) and higher yields in phosphogypsum treated fields. Greater increase in aggregation, soil organic carbon, microbial biomass carbon, aggregate associated carbon, hydraulic conductivity and moisture retention capacity were observed in phosphogypsum treated soils.

Since phosphogypsum is a byproduct of natural rock, some contaminants in the rock are likely to affect its quality. It is assessed that application of 10 Mg ha^{-1} PG is likely to add about 10.7 kg F ha^{-1}. Besides danger posed by radioactive nuclei is also anticipated. These issues have been addressed in the studies on phosphogypsum application as an amendment. Higher soluble F in PG treated fields was observed compared to mineral gypsum treated fields. The F content in rice and wheat straw and grains were also higher than mineral gypsum but came to about the same level in the 3rd year and afterwards. The fluoride content in straw was found to be much less than the zoo toxic threshold in herbage, which is reported to be 30 mg kg^{-1}. The results suggest that one time application of PG @ 10.0 t ha^{-1} as an amendment to alkali soil may not pose zoo toxicity threat.

Table 4.7 Effect of phosphogypsum on ESP of soil after first and third year of rice-wheat cropping

Treatments	2006-07		2008-09	
	0-15 cm	15-30 cm	0-15 cm	15-30 cm
50 GRm	26.5	31.6	24.6	28.5
50 GRpg	20.4	27.3	18.3	25.7
50 GR_{PG}-KFC	18.3	25.2	17.9	23.4
25 GRm+ 25 GRpg	17.8	23.5	16.7	21.5
25 GRpg	19.4	25.6	17.9	24.3
12.5 GRm+ 12.5 GRpg	25.2	30.8	23.5	27.5
LSD (p=0.05)	1.5	2	1.5	2

50GRm = 50% gypsum requirement through mineral gypsum; 50 GRpg = 50% gypsum requirement through phosphogypsum; 25 and 12.5 means 25% and 12.5% of GR, KFC means application followed by cultivation of Karnal grass

The PG used in the trials contained 1895 ± 39 to 2120 ± 56 Bq kg^{-1} of ^{226}Ra. Radio nuclides analysis showed that no detectable levels of radio nuclides were found in the vegetal tissues such as grain and straw of rice and wheat grown on alkali soil with PG application. In the European community, unconditional release limit for the material containing radium is 10000 Bq kg^{-1}. Material below this level may be used for any purpose and phosphogypsum is used in wall board, cement and as a soil conditioner. It seems that one time application of the PG may not result in detectable radio nuclides in rice and wheat.

Cultural Practices for Rice-Wheat Cropping

Transplanted rice-wheat has emerged as a standard cropping sequence to reclaim alkali lands. Although Rice-*shaftal* has also shown good promise but the former cropping sequence continues to be popular. Similarly, the reclamation process beginning with *kharif* crops other than rice didn't prove attractive to the stakeholders. Therefore, major attempts have been made to finalize cultural practices to cultivate rice and wheat crops that help to get higher yields even during the initial years of reclamation. A look at these practices seems to be in order for this chapter.

Rice

Practice	What to do
Seedling preparation	Prepare healthy seedlings on a normal soil
Field preparation	The infiltration rate of these soils being low and weeds may not create many problems either; avoid puddling for first 2-3 years.
Age of seedlings	Transplant 35-40 days rather than normally recommended 30 days old seedlings
Number of seedlings	To take care of high mortality and less tillering, use higher number of seedlings per ha. Use row to row and plant to plant spacing of 15×15 cm and 4-6 seedlings per hill to increase the number of seedlings per ha.
Nutrient management	Apply 25% more nitrogen than the recommended dose for normal soils in 3 splits
	Apply phosphorus and potash as per soil tests. At many places in Haryana and Punjab, these soil are high in P and K such that one can skip the application of these nutrients for several years until their deficiency is revealed by soil tests
	Application of zinc as $ZnSO_4$ is essential for getting high yields in these soils. Apply 10-20 kg Zn ha^{-1} every year to rice crop.
Integrated nutrient management	Cultivate *Sesbania* (*Dhaincha*) after wheat to green manure the fields before rice. Plough the crop when it is 40-45 days old, and allow it to decompose for about 7 days under anaerobic conditions before transplanting rice crop. If time is a constraint, transplant rice as soon as possible since decomposition will occur under standing water in the rice crop. Nitrogen dose in that event can be reduced accordingly.
	Apply FYM, press mud or follow crop residue management practices to improve organic matter content of the soils

Wheat

Practice	What to do
Seed rate	Use 20-25% higher seed rate than recommended for normal lands
Time of first irrigation and numbers of irrigation	Delay first irrigation by about a week or so compared to normal lands and later apply shallow and frequent irrigations, total water applied remaining the same apply 5-6 instead of 4-5 irrigation in normal lands
Hoeing	To ensure proper aeration and avoid crust formation, hoeing after first and second irrigation proved useful

Practice	What to do
Nutrient management	Apply 25% more nitrogen than the recommended dose for normal soils in 3 splits
	Apply phosphorus and potash as per soil tests. At many places in Haryana and Punjab, these soil are high in P and K such that one can skip the application of these nutrients for several years until their deficiency is revealed by soil tests
	Application of zinc as $ZnSO_4$ may not be essential if applied to rice crop @ 20 kg ha^{-1}. As a precaution, apply $ZnSO_4$ @10 kg Zn ha^{-1} for the first 2-3 years to wheat crop as well
Integrated nutrient management	Cultivate *Sesbania* (*Dhaincha*) after wheat to green manure the fields before rice.
	Apply, FYM or follow residue management practices to improve organic matter content of the soils.

Saline-alkali Soils

Most alkali soils in India are saline-alkali in nature, both salinity (EC_e more than 4.0 dS m^{-1} in most cases and 2.0 dS m^{-1} in all cases) and ESP being high. From management point of view, saline-alkali soils are treated as alkali soils. As per the international categorization, it may appear odd but practically it is logical. Nonetheless, there are some finer points that help to reduce the cost of reclamation of saline-alkali soils as well as enhance the pace of reclamation.

Soil EC_e in alkali soils is due to the presence of excessive carbonates and bicarbonates of sodium. As such when GR is determined using the Schoonover method, estimated GR is quite high. As already explained, in such cases, it pays to work out the GR using Abrol et al. (1975) method. Another way is to apply a heavy pre-sowing irrigation to leach down the salts before drawing soil samples. But this practice may pose problems in heavy textured highly saline soils as experienced in central India. Sampling after flushing in that event is a better alternative.

Flushing or leaching of the salts from saline-alkali soils has proved to be useful practice especially in fine textured highly saline alkali soils having saline efflorescence at the soil surface. Two flushing with an interval of about 7 days is recommended. For effective flushing, water is kept ponded on the land surface for 2-3 days. It is followed by puddling to dissolve soluble salts. Water is then drained through the surface drainage system. Since the soluble carbonates and bicarbonates of sodium are effectively flushed, gypsum application following flushing proved more effective than without flushing. Compared with conventional treatment, the surface flushing significantly reduced the EC_e and SAR of the soil and improved physical properties such as f-potential, dispersible clay content, water stable aggregates and saturated hydraulic conductivity (Nayak *et al.*, 2008). Split application of gypsum between flushing was not beneficial and therefore, full dose of gypsum should be applied after second flush. With flushing average yields of rice and wheat increased by 25% and 62%, respectively as compared to the conventional practice of leaching alone.

Land forming may not play a significant role in managing alkali soils. But in saline-alkali soils, it does help to manage these soils and can be adopted as a management practice. A trial on sugar beet on saline-alkali soil (initial pH_2 10.5 and EC_e 6.36 dS m^{-1}), sowing the seeds on one side of the ridge halfway between the top and bottom opposite to the sun direction gave significantly higher yield than in flat method. Thus, land forming practices discussed in Chapter 3 can be employed for saline-alkali soils wherever feasible.

■■■■

Chapter - 5

ROLE OF CROPS IN MANAGEMENT OF SALT AFFECTED SOILS

All the cultivated plant species are glycophytes including facultative halophytes. These are unable to tolerate salt stress beyond some permissible levels. Therefore, when cultivated under stressed environments most show reduced growth and yields. The degree of adverse effects varies depending on the species, strains, phase of development, the concentration and composition of salts present in the root medium and the plant's exposure to water regime. Since the complete reclamation of salt affected soils especially saline soils is time consuming, scientists have been looking at plants that can adapt to salt affected soils. Besides, attempts have also been made to reduce the cost of land reclamation by growing relatively salt tolerant plants, as it may not always be cost-effective to fully reclaim the soils. Fortunately, we have a reasonably good continuum of salt tolerance in plants. On one hand, we have 'glycophytes', the plants very sensitive to salts say chickpea and on the other we have halophytic plants that are highly tolerant to salts, say *Tecticornia.* Cultivation of salt tolerant species and strains has emerged as an effective means to manage salt affected soils. This chapter illustrates the role of plants in diagnosis of salt affected soils and role of selection of crops and varieties as a tool to manage these soils.

INDICATOR PLANTS

Some kind of relationship between plant associations, the water table depth and soil and groundwater salinity has been observed. As such certain native plants can serve as valuable indicators in the diagnosis of salt affected soils. Although in USA, work began as early as at the beginning of the 20th century (Hilgard, 1906), yet not much work in India was done until the 4th decade of the century (Hoon and Mehta, 1937, Mehta, 1951). Agarwal et al. (1982) reviewed this and other works, which is summarized as Table 5.1.

In an attempt to devise visual diagnostic tools to distinguish saline and alkali soils, indicator grasses and plants for theses soils were identified. The respective grasses for saline and alkali soils proved to be a good diagnostic criterion.

In highly saline soils, following grasses come up during and after the monsoon season, when soil salinity is low as a result of leaching from rainfall.

Cressa cretica, Cyperus rotundus, Chloris pallida, Sporobolus pallidus, Haloxylon salicornicum, Acluropus lagopoides, Zygophyllum simplex, Dichanthium annulatum, Sueda fruticosa, Butea monosperma.

In highly alkali soils, following grasses come up during and following the monsoon season, when soil salinity is subdued although there may not be much change in ESP.

Sporobolus marginatus, Desmostachya bipinnata, Suaeda maritma, Diplachne fusca, Dichanthium annulatum, Cynodon dactylon, Chloris barbata, Brachiaria mutica, Kochia indica, Panicum antidotale.

Table 5.1 Some of the indicator plants of salt affected soils

Indicator plant	Land status
Wan (*Salvadora oleoides*)	Moderately alkaline
Okan (*Tamarix articulate*)	Saline throughout the profile
Karai (*Capparis aphylla*)	High exchangeable sodium or pH
Aak (*Calotropis procera*)	Moderate alkalinity
Sarkanda (*Saccharum munja*)	Wet soils or soils with shallow water table
Lani (*Suaeda fruticosa*)	Soil salinity
Dhak (*Butea monosperma*)	Advanced stage of soil degradation
Khar usar (*Sporobolus arabicus*)	Advanced stage of soil degradation
Usranta (*Chloris variegata*)	Alkali soils
Dab or kusa (*Desmostachya bipinnata*)	Better pockets of saline-alkali soils
Dub (*Cynodon dactylon*)	Better pockets of saline-alkali soils
Janewa (*Dichanthium annulatum*)	Better pockets of saline-alkali soils
Gantar (*Andropogon squarrosus*)	Better pockets of saline-alkali soils
Salicornia sp.	Coastal salinity

In a survey Gururaja Rao et al. (2001) studied the soil characteristics under several plant species in various tracts of salt affected regions of Gujarat.

On highly saline soils of Kharad (EC_e, 17 dS m^{-1}) and Kalamsar, Khanpur and Saragwada (EC_e,70-132 dS m^{-1}), plant species like *Cressa cretica, Capparis decidua, Prosopis juliflora, Salvadora persica.* S. *oleoides, Suaeda maritima,* S. *nudiflora, Tamarix ericoides* and grass species like *Aleuropus lagopoides, Cynodon dactylon, Sporobolus airoides and Eragrostis* were observed. On moderately saline soils where soil salinity ranged from 1.25 to 8.0 dS m^{-1}, plants like *Acacia nilotica, Agave americana, Ailanthus excels, Albizia lebbeck, Asparagus racemosus, Borassus flabellifer, Cassia auriculata, Prosopis cineraria,* and grass *Dichanthium annulatum* were growing.

The reduced soil salinity after the monsoon season has a great bearing on the natural vegetation. The prominent species noticed after the monsoon are: *Chenopodium album, Digera muricata, Heliotropium supinum, Sphaeranthus indicus, Celosia argentia, Boerhavia dijfusa, Catharanthus roseus* and grasses *Eragrostis ciliaris, Eulalia trispicata, Dactyloctenium aegyptium, Aristida adscensionis* and some sedges. These have great economic value to the local people, being used as forage for the cattle. Occurrence of these plant species after monsoon and their non-occurrence during pre-monsoon period indicates their susceptibility to high salinity prevailing during summer months.

The dominant plant species, occurring both during pre- and post-monsoon, followed by co-dominant ones seen frequently on saline soils are reported in Table 5.2. Thus the relative distribution indicates that *Prosopis juliflora, Salvadora perisca, Tamarix ericoides, Cressa cretica, Suaeda maritima, Suaeda nudiflora, Aleuropus lagopoides, Cynadon dactylon* are the potential candidates for further exploitation.

Table 5.2 Preponderance of plant species in saline soils in different areas during summer

Dominant	*Prosopis juliflora, Salvadora perisca, Tamarix ericoides, Cressa cretica, Suaeda maritima, Suaeda nudiflora, Aleuropus lagopoides, Cynodon dactylon*
Co-dominant	*Prosopis cineraria, Capparis decidua, Sporobolus airoides*
Frequent	*Pithecellobium dulce, Zizyphus nummularia, Dichanthium annulatum, Eragrostis* species
Occasional	*Acacia nilotica, Albizia lebbeck, Albizia procera, Salvadora oleoides, Borassus flabellifer, Dendrocalamus strictus, Parkinsonia aculeata, Asparagus racemosus, Calotropis gigantea, Heliotropium supinum, Sphaerathus indicus*
Rare	*Ailanthus excelsa, Azadirachta indica, Tamarindus indicus, Aregemone mexicana, Boerhavia diffusa, Caesulaia axillaris, Cassia auriculata, Cheneopodium album, Digera muricata*

Jagtap et al. (2002) studied the vegetative structure (distribution and growth pattern, density, and biomass) and physico-chemical characteristics of the ambient environment of the major *Salicornia* and *S. brachiata* beds along the Gulf of Kutch (northwest) and Tamil Nadu (southeast) coasts. Some *Salicornia* species are able to tolerate nearly twice the salinity of seawater. It was observed that the width of the beds varied from approximately 5 m to 1000 m, depending upon the topography and tidal influence. The most extensive and healthy stands occurred in the regularly (20–50 tides month^{-1}) inundated intertidal zones. The width of *Salicornia* beds in the Gulf of Kutch was greater than those along the Tamil Nadu coast due to greater (0.83–7.2 m) tidal amplitude and flat topography. The growth of the *S. brachiata* seemed to be influenced primarily by the frequency of inundation, duration of atmospheric exposure at ebb tide, rainfall, and salinity and salt content of the ambient water and sediments, respectively. While *Salicornia* may be a good indicator of coastal salinity, its distribution and general health may be affected by other factors, for which necessary indices need to be developed.

Arora and Mehta (2018) surveyed the natural vegetation in the bara tract of Gujarat. They identified plants that were growing under a wide range of pH (8.33 to 9.55) and EC_e (0.55 to 82 dS m^{-1}) in the 0-50 cm depth (Table 5.3).

Table 5.3 Rhizospheric properties of soil under some salt tolerant plant species and halophytes

Plant species	pH	EC_e (dS m^{-1})
Cressa cretica	8.43	0.55
Suadea maritima	9.38	3.66
Suadea nudiflora	8.35	30.00
Salvadora persica	8.84	9.41
Atriplex nummularia	8.65	15.50
Prosopis cineraria	8.33	35.50
Acacia nilotica	9.40	2.00

No doubt, plants are good indicators of various kinds of problems but there are few limitations that require careful use of this approach so as to draw valid conclusions.

- Most indicator plants that can grow under stressed environment can also grow well under normal soils. Thus, one has to be quite cautious while drawing any conclusion. It may be worthwhile to develop quantitative methods that relate the growth performance of indicator plants viz-a-viz their ability to survive under the stressed environment as revealed through the physical and chemical measurements of the soils.
- The second limitation may arise due to the soil sampling. There are large spatio-temporal variations in different parameters within a small distance around the plant canopy. The physical and chemical parameter may also be quite different if taken before or after the monsoon season. Therefore, one has to very careful in planning the sampling strategy and the number of samples to be drawn from each location. All relevant parameters that might impact the plants growth should be assessed.
- It may be only by chance that one may encounter pure stands. Mostly plant associations are mixed. Vegetative growth and vigour may vary from one plant species to another. Therefore, observations have to be used in relation to growth of all dominant species. It makes the task even more complex to arrive at any worthwhile conclusion.

Crop Condition as a Diagnostic Criterion

As a result of large spatio-temporal variation in soil salinity, crop/plants growing on salt affected soil usually show the following symptoms.

- Patchy growth (spots having good growth at few locations but no growth at others in the same field)
- Stunted plant growth
- Variable vigour and size

- Deep blue-green foliage
- Chlorosis

It may be noted that these features or symptoms are not exclusive to salt affected soils. These features may be noticed in improperly levelled fields, in fields having nutrient deficiencies and inadequacy of water. Similarly, chlorosis may be due to calcareous nature of the soil. In some cases, it may also show up when crops are irrigated with water having high bicarbonate content. Some species of plants develop characteristic necrotic areas and tip burn and firing of the margins of the leaves. Even these symptoms may appear from other factors that cause malfunction of the root system, such as root diseases and subsurface water logging. These symptoms may be indicative of salinity only when they occur in the presence of apparently adequate moisture and nutrient status and disease free condition. Nonetheless, these symptoms, if carefully evaluated discounting other problems, may give good indication of soil salinity.

INTRA CROP VARIABILITY IN SALT TOLERANCE

Most cultivated plants are known glycophytes but fortunately have a large range of inter and intra-crop variability for salt tolerance (Table 5.4). While rice has the maximum range of genetic variability for alkalinity, barley has the maximum for salinity tolerance. This variability, besides its role in crop selection, reflects on the future success of any breeding program for stressed environments.

Table 5.4 Intra-crop variability of salt tolerance in some important crops

Crop	Alkalinity (pH)	Salinity (EC_e, dS m^{-1})
Rice	9.2-10.2	4.0-10.0
Wheat	8.8-9.3	6.5-11.0
Barley	9.0-9.3	4.0-11.0
Indian mustard	8.9-9.2	4.0-11.0
Toria	8.8-9.0	4.0-6.0
Taramira	8.7-8.9	4.0-6.0
Linseed	8.7-8.9	4.0-6.0
Safflower	8.7-8.8	4.0-6.0
Sunflower	8.6-8.8	4.0-6.0
Sugarcane	8.8-9.0	-
Cowpea	8.6-8.8	4.0-5.0
Green gram	8.6-8.8	-
Pigeon pea	8.3-8.6	-
Pea	8.6-8.8	4.0-5.0
Chick pea	8.6-8.8	4.0-5.0
Bakla	8.6-8.8	4.0-7.0
Sugar beet	9.2-9.8	4.2-10.0
Pearl millet	8.7-8.9	4.0-8.0

Relative tolerance of crops is often judged using any of the following four criteria.

- Germination of crops
- Plant survival
- Absolute plant growth or yield
- Relative growth or yield

Germination

Large numbers of investigators have tested the salt tolerance of crops at germination stage and attempts made to extrapolate the tolerance limit for final performance (Sinha et al., 2003). More than often, such attempts failed to give satisfactory results because tolerance characteristics differ from one stage of growth to another. Nonetheless, most crops are sensitive to salts at the germination stage. For example sugar beet known to be highly tolerant to salinity, is quite sensitive to salinity during germination stage but is quite salt tolerant at later growth stages. Other notable crops that have low salt tolerance at germination are barley, wheat, alfalfa and many others. To illustrate, wheat for example is least tolerant to salts at crown root initiation followed by tillering, jointing, booting, milking and dough in that order. Moreover, salts accumulate in the top few centimeters of soil profile where seeds reside. High salinity of this layer at the time of germination results in delayed or poor germination. The poor crops stand resulting from inadequate germination is reflected in poor yields. Therefore, one of the ways to get good germination is to leach the salt out as per the salt tolerance of the crop at the germination stage. Data on salt tolerance at germination stage can prove to be handy for this purpose. Few other practices to improve germination are discussed in the following paragraphs.

Besides the germination, the inhospitable soil environment results in mortality of young seedlings. Coupled with poor tillering, crops yields are generally poor. To counter these effects a high seed rate is used as a management option. For cotton, pearl millet, sorghum, wheat, barley, mustard etc. about 25% higher seed rate over the recommended seed rate for normal soils ensures good crop stand. In case of transplanted crops, the number of seedlings per hill should be increased and/or plant to plant and row to row distance should be decreased.

In saline soils, it is also possible to avoid the salt accumulation around the seed through land modification and planting practices discussed in Chapter 3. Several practices are described that help to improve the crop stand of even sensitive crops.

Chemical priming of seeds has been found to improve salt tolerance of crops at the germination stage. In a field experiment, seed treatment with 3% Na_2SO_4 resulted in higher yield of wheat in a saline soil (pH 7.6, EC_e 10.5 dS m^{-1}). Compared to control, wheat yield was 3 times in S-227 and two times in K-65 varieties respectively (Puntamkar *et al.*, 1971). In another study, it was observed that grain yield of wheat and barley crops were not significantly affected when seeds were pre-soaked in NaCl or Na_2SO_4 salt solutions but pre-soaking in low salinity tube well water gave higher grain yield in both the crops.

Pre-soaking of seeds in hormone solutions @ 50, 100, 200 and 300 mg L^{-1} of α-naphthaleneacetic acid (NAA), indole-3-acetic acid (IAA), indole-3-butyric acid (IBA) and gibberellic acid (GA_3) was tested to assess its effect on the growth and development of wheat (S-227) irrigated with saline water (EC_{iw} 12 dS m^{-1}, SAR 30, boron 2 and 4 mg L^{-1}). The results revealed that phytohormones generally improved the growth but caused differential effect in plant height, root length, shoot growth and grain yield. The hormones produced significant positive changes in various growth attributes up to 200 mg L^{-1} but adverse effects appeared at 300 mg L^{-1}.

Germination of Indian mustard (*Brassica juncea* (L.) Czernj & Cosson) was 100% in water soaked non-chilled (control) seeds up to 100 mM NaCl, but went up to 200 m*M* NaCl in chilling treated seeds. Pre-sowing chilling treatments of seeds for 5, 10 and 15 days also enhanced the dry mass of 6-d-old seedlings, concentrations of saccharides and amino acids. The alleviation of NaCl stress by pre-sowing chilling treatments was also associated with decrease in Na^+ and proline accumulation and a slight increase in K^+ and Ca^{2+} contents of seedlings (Sharma and Kumar, 1999).

Although the positive impacts of seed treatment are visible yet it seems that much more efforts are needed to firm up sound recommendations on this issue.

Plant Survival and Absolute Yields

Plant survival has been used as a criterion by ecologists but without yield, plant survival alone has little value to the farmer. On the other hand, it can prove to be useful criterion for plant breeders.

Absolute plant growth or yield is of greatest interest to the farmers. However, this criterion does not permit comparison between crops because yields of different crops are not expressed in comparable terms.

Relative Yield

Relative growth or yield, defined as the yield on a saline/alkali soil as a fraction of the yield on a non- saline/alkali soil under similar growing environment and nutritional conditions, seems to be the most desirable and popular criteria used in crop tolerance studies. The main difficulty with this method is to decide at what level of yield reduction plant's relative tolerance should be compared? A researcher comparing the salt tolerance at 10% yield reduction may essentially end up differently than the other who is comparing the salt tolerance at 50% yield reduction. Nevertheless, this criterion is favoured world over and is widely used for relative comparison. Normally, 50% yield reduction is used to prepare relative salt tolerance tables by the scientific community. For the farmers, probably a 10% yield reduction criterion should be adopted since yield reduction beyond this value may have economic implications.

RELATIVE TOLERANCE OF CROPS

Crops differ widely in their tolerance to saline/alkali environment of the root zone. Even crop varieties respond differently to these conditions. Although not very extensive, but salt tolerances of many species and varieties of crop plants have been investigated at CSSRI, Karnal. These data sets along with collated information from various sources are used in India to take knowledgeable decision on crops and varietal selection. It is now known that tolerance of crops for the saline and alkali stresses will not be similar mainly because the causes of yield reduction in both the environments is due to different reasons as below.

Alkali soil	Saline soil
• Structural deterioration of the soil (adverse physical condition) restricting moisture transmission and aeration. It may physically impede root elongation and seedling emergence. • Nutritional imbalance due to high pH and high concentrations of HCO_3 and CO_3 or excess of Na causing imbalance in the uptake of other ions especially K and Ca. • Direct toxic effects of Na, HCO_3 and CO_3 ions	• A general osmotic effect in which growth and yield are determined by the osmotic pressure of the medium. • Toxic effects on specific plants due to accumulation of the causative ions in the plant e.g. Na^+, Cl^-, SO_4^{--}, Mg^{2+}, B. • Imbalance in the uptake of ions • High salinity may affect several metabolic processes such as CO_2 assimilation, protein synthesis and respiration

Therefore, all the information presented in this chapter is categorized under the two heads namely soil salinity and crop tolerance and soil alkali and crop tolerance. Besides, varietal differences in salt tolerance are discussed under a separate head.

Soil Salinity and Crop Tolerance

The widely used tabulated information on relative tolerance to soil salinity is reported in Table 5.5. It classifies the soil in sensitive and tolerant classes. Within these classes, crops have further been categorized in two groups each making a total of four categories. Data base reported by Maas and Hoffman (1977) is also widely used in India. According to their formulation, crop yield has been described by a piecewise linear model expressed as:

$$Y = 100 \qquad 0 \leq EC_e \leq EC_t \qquad (5.1)$$

$$Y = 100 - S(EC_e - EC_t) \qquad EC_t \leq EC_e \leq EC_0 \qquad (5.2)$$

$$Y = 0 \qquad EC_e \geq EC_0 \qquad (5.3)$$

Here Y is the relative yield of the crop, %, EC_e is the electrical conductivity of the saturation extract of the soil, EC_t is the threshold EC_e up to which there is no decline in the yield of the crop, S is the slope that represents the yield reduction per unit increase in the EC_e of the soil solution beyond the threshold EC and EC_0 is the EC_e at or beyond which crop fails to give any yield. Eq. (5.2) is commonly used to determine the yield in between EC_t and EC_0. For the Indian conditions, the threshold EC and EC_e for which relative yield is 50% of the yield on a non-saline soil for few agro-climatic conditions are given in Table 5.6. The variations in tolerance are attributed to soil types and climatic conditions besides the management practices adopted during cultivation. Mass (1986), data are also based upon this model and has a greater application potential (Fig. 5.1 and Table 5.7). It may not be out of place to mention that these tables are based on a field-plot technique in which crops are grown on soils that are artificially adjusted to various salinity levels after the seedlings are established. Thus, any change in management may change the relative tolerance of crops. Clearly, these data sets only provide some basic guidelines although the actual tolerance may vary depending upon the management.

Table 5.5 Crop groups based on response to salt stress

Sensitive groups		Resistant groups	
Highly sensitive	Medium sensitive	Medium tolerant	Highly tolerant
Lentil	Radish	Spinach	Barley
Mash	Cowpea	Sugarcane	Rice (transplanted)
Chickpea	Broad bean	Indian mustard	Cotton
Beans	Vetch	Rice (direct sowing)	Sugar beet
Peas	Cabbage	Wheat	Turnip
Carrot	Cauliflower	Pearl millet	Tobacco
Onion	Cucumber	Oats	Safflower
Lemon	Gourds	Alfalfa	*Taramira*
Orange	Tomato	Blue panic grass	Karnal grass
Grape	Sweet potato	Para grass	Date palm
Peach	Sorghum	Rhodes grass	*Ber*
Plum	Minor millets	Sudan grass	Mesquite
Pear	Maize	Guava	*Casurina*
Apple	Clover, *berseem*	Pomegranate	*Tamarix*
		Acacia	*Salvadora*

Table 5.6 Crop tolerance to soil salinity

Sites	Soil type	Crop	Threshold EC_t ($dS\ m^{-1}$)	Slope (%)	EC_{50} ($dS\ m^{-1}$)
Sample (Haryana)	Sandy loam	Wheat	4.0	29.0	5.7
		Mustard	6.0	15.0	9.3
		Barley	7.0	19.0	9.6
Karnal (Haryana)	Sandy loam	Green gram	1.8	20.7	3.3
		Mustard	3.8	6.9	11.0
		Sorghum	2.2	10.6	6.9
Agra (Uttar Pradesh)	Sandy loam	Wheat	8.2	19.8	10.7
		Mustard	6.1	20.7	8.5
		Berseem	3.5	12.5	7.5
		Tomato	1.3	6.5	9.0
Dharwad (Karnataka)	Black clay	Wheat	2.3	20.5	4.7
		Safflower	2.8	20.7	5.2
		Sorghum	2.1	3.9	14.9
		(winter)	6.1	50.0	7.1
		Italian millet			
Indore (Madhya Preadesh)	Black clay	*Berseem*	2.3	20.5	4.7
		Safflower	2.8	20.7	5.2
		Sorghum	2.1	3.9	14.9
		(winter)	6.1	50.0	7.1
		Italian millet	1.1	2.2	23.9
		Sataria			

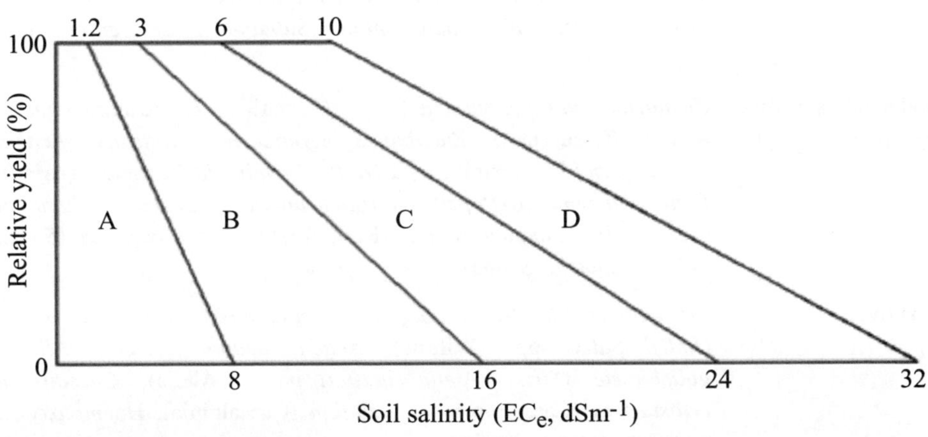

Fig. 5.1 Crop classification based on categorization proposed by Maas (1980)

Tolerance of agro-forestry plants to soil salinity listed in Table 5.8 (Tomar and Gupta, 1986) reveal that most tolerant plants are: *Tamarix troupii* (*Pharans*), *T. articulata* (*Pharans*), *Prosopis julifora* (*Pahari Kikar*), *Pithecellobium dulce* (*Junglee jalebi*), *Parkinsonia aculeata* (Parkinsonia), *Acacia farnesiana* (Pissi

Babul). When salinity is not associated with shallow water table, *Acacia auriculifarmis, Leucaena leucocephala* can be grown. Where high salinity and shallow water table are encountered, species like *Casuarina glauce, Tamarix articulata* and *Prosopis juliflora* perform well.

Table 5.7 Salt tolerance of some important agricultural crops (Mass, 1986)

Salt tolerance as per Fig. 8.2	Crops
A, sensitive	Chickpea, Beans, Sesame, Onion, Carrot, Green gram
B, moderately sensitive	Broad bean, Maize, Groundnut, Sugarcane, Alfalfa, *Berseem*, Radish, Paddy, Cauliflower, Chilly
C, moderately tolerant	Wheat, Mustard, Cowpea, Oats, Sorghum, Soybean, Spinach, Pearl millet, Muskmelon
D, tolerant	Barley, Cotton, Sugar beet, Safflower

Table 5.8 Salinity tolerance of trees

Tolerant (EC_e 25-35, dS m^{-1})	*Tamarix troupii* (*Pharans*), *T. articulata* (*Pharans*), *Prosopis julifora* (*Pahari Kikar*), *Pithecellobium dulce* (*Junglee jalebi*), *Parkinsonia aculeata* (Parkinsonia), *Acacia farnesiana* (*Pissi Babul*)
Moderately tolerant (EC_e 15-25, dS m^{-1})	*Callistemon lancheolatus* (Bottle brush), *Acacia nilotica* (*Kikar*), *A. pennatula* (*Babul*), *A. tortilis* (*Israeli kikar*), *Casuarina glauca 13155* (Casuarina), *C. glauca 13987* (Casuarina), *C. obesa 27, C. glauca* (FRI), *C. equisetifolia* (FRI) (Casuarina), *Eucalyptus camaldulensis* (*Safeda*), *Leucaena leucocephala* (*Subabool*), *Erescentia alata* (Erescentia)
Moderately sensitive (EC_e 10-15, dS m^{-1})	*Casuarina cunninghamiana* (FRI) (Casuarina), *C. cunninghamiana* (Aust) (Casuarina), *Eucalyptus tereticornis* (*Safeda*), *Acacia auriculiformis* (Australian *kikar*), *Terminalia arjuna* (Arjan), *Pongamia pinnata* (*Papri*), *Gauzuma ulmifolia, Leucaena shannoni* (*Subabool*), *Samanea saman* (Rain tree), *Albizia cartibea* (*Siris*), *Senna atomaria* (Senna)
Sensitive (EC_e 7-10, , dS m^{-1})	*Syzygium cumini* (*Jamun*), *S. fruticosum* (*Jamoa*), *Tamarindus indica* (*Imli*), *Salix* spp. (Willow), *Acacia deami* (Acacia), *Albizia guachapele* (*Siris*), *Alelia herbertismithi* (Alelia), *Caesalpinia eriostachya* (Caesalpinia), *C. velutina* (Caesalpinia), *Haemotoxylon brasiletto* (Haematoxylon)

Soil Alkalinity and Crop Tolerance

Pearson (1960) listed the crops as per their tolerance to soil ESP (Table 5.9). Using the data from numerous studies conducted at CSSRI, Karnal, Gupta and Sharma (1990) derived information on crop tolerance to soils ESP. They modified and used the piecewise linear model of Maas and Hoffman (1977) to describe the relative alkali tolerance of crops (Table 5.10). The threshold ESP, ESP at 50% yield reduction and slope (% yield reduction per unit increase in ESP beyond the threshold) were derived and are reported in Table 5.10. Rice and wheat turns out to be quite tolerant having high threshold ESP_t. The relative crop tolerance table widely used in India is reported as Table 5.11. It may be emphasized that like crop tolerance to soil salinity, crop tolerance to soil alkalinity also vary with the stage of growth, variety, environmental factors and level of management.

Table 5.9 Tolerance of various crops to exchangeable sodium (ESP) under non-saline conditions

Tolerance to ESP	Crops	Growth response under field conditions
Extremely sensitive (ESP = 2-10)	Deciduous fruits, Nuts, Citrus (*Citrus* spp.), Avocado (*Persea americana* Mill.)	Sodium toxicity symptoms even at low ESP values
Sensitive (ESP - 10-20)	Beans (*Phaseolus vulgaris* L.)	Stunted growth at these ESP values even though the physical condition of the soil is not affected much
Moderately tolerant (ESP - 20-40)	Clover (*Trifolium* spp.) Oats (*Avena saliva* L.) Tall fescue (*Festuca arundinacea* Schreb.) Rice (*Oryza saliva* L.) Dallis grass (*Paspalum dilatum* Poir.)	Stunted growth due to both nutritional factors and adverse soil conditions
Tolerant (ESP - 40-60)	Wheat (*Triticum aestivum* L.) Cotton (*Gossypium hirsutum* L.) Alfalfa (*Medicago sativa* L.) Barley (*Hordeum vulgare* L.) Tomatoes (*Lycopersicon esculentum* Mill.) Beet, garden (*Beta vulgaris* L.)	Stunted growth usually due to adverse physical conditions of soil
Most tolerant (ESP more than 60)	Crested and Fairway wheatgrass (*Agropyron* spp.) Tall wheatgrass (*Agropyron elongatum* Host Beau.) Rhodes grass (*Chloris gayana* Kunth.)	Stunted growth usually due to adverse physical conditions of soil.

Table 5.10 Threshold, slope and ESP_{50} for alkali tolerance of crops for medium textured soils (sandy loam)

Crop	Botanical name	Piecewise linear model		
		ESP_t	(%) Slope	ESP_{50}
Barley	*Hordeum vulgare* (P)	8.5	3.5	22.3
Cowpea	*Vigna unguiculata*	13.5	9.1	19.0
Sesbania	*Sesbania aculeata*	46.9	2.4	67.7
Gram	*Cicer arietinum*	7.7	5.0	17.7
Groundnut (Peanut)	*Arachis hypogaea*	8.0	2.2	29.7
Guar (Cluster bean)	*Cyamopsis psoraloides* D.C.	11.9	3.2	27.5
Lentill	*Lens esculentum*	4.9	5.5	14.0
Linseed	*Linum unitatissinum* L.	13.3	2.3	25.0
Pear millet	*Pennisetum typhoideum*	13.6	2.6	32.8
Raya (Indian mustard)	*Brassica juncea* L.	7.6	0.8	70.1
Rice	*Oryza sativa*	24.4	0.9	80.0+
		32.7	1.3	71.2*
Rye	*Secale cereale*	11.0	2.6	30.2
Sesamum	*Sesamum indicum*	9.0	3.6	22.8
Soybean	*Glycine max*	8.0	3.5	22.3
Sunflower	*Helianthis annus* L.	11.3	1.1	56.8
Saffflower	*Carthamus tinctorius*	7.6	5.2	17.2
Wheat	*Triticum aestivum* (P)	16.4	2.1	40.2+
		20.4	2.6	39.6*
Pea	*Pisum sativum*	7.7	4.1	19.9
Onion	*Allium cepa*	9.8	2.2	32.5
Garlic	*Allium sativum*	9.5	1.8	37.3

ESP values indicate average soil ESP of 0-15 cm layer, data from pot (*) and field (+) experiments

Table 5.11 Tolerance of crops to various levels of ESP

Range of ESP	Crops
10 – 15	Safflower, mash, peas, lentil, pigeon pea, *urd* bean
15 –20	Bengal gram, soybean, maize
20 – 25	Groundnut, cow pea, onion, pearl millet
25 – 30	Linseed, garlic, guar, lemon grass, pamarosa, sugarcane, cotton
30 – 50	Wheat, *raya*, sorghum, sunflower, *berseem*, *shaftal*, *senji*, blue panic, gutton panic
50 – 60	Barley, *sesbania*, para grass, Rhodes grass, matricaria
60 - 70	Rice, sugar beet, Karnal grass

Note: Relative crop yields are only 50% of the maximum in the indicated ESP range

Central Soil Salinity Research Institute has made significant contribution in testing various grasses and forage crops (Table 5.12). Karnal grass, Rhodes grass, Para grass and Bermuda grass are highly tolerant and can be successfully grown in alkali soils (Kumar and Abrol, 1986). Karnal grass grows extremely well in soils of very high ESP (80 to 90) even without application of any amendment. In heavy textured Verisols highly salt tolerant and high biomass producing grass species besides above mentioned species include *Aeluropus lagopoides*, *Sporobolus helvolus*, and *Brachiaria ramose*. Information on agro-forestry and fruit plants for their tolerance to alkali environment is also generated (Table 5.13). The most promising species for planting in soils having pH more than 10.0 include: *Pahari Kiker* (*Prosopis juliflora*), *babul* (*Acacia nilotica*), *Frans* (*Tamarix articulata*) and Australian pine (*Casuarina equisetifolia*). *Salvadora persica*, a facultative halophyte appears to be a potentially valuable oil seed crop for saline and alkali soils. Its seeds contain 40–45% of oil rich in industrially important lauric (C12) and myrestic (C14) acids. The most promising fruit species are: *Aonla* (*Emblica officinalis*), *Karaunda* (*Carissa carandus*) and Guava (*Psydium guajava*) (Tomar et al., 1998, Singh et al., 1993; Dagar et al., 2001).

Table 5.12 Promising forage grasses/crops for alkali soils

pH	Grass/forage crop (Botanical name)
Grasses	
10.4	Karnal grass (*Leptochloa fusca*), Rhodes grass (*Chloris gayana*), Gatton panic (*Panicum maximum*), Bermuda grass (*Cynodon dactylon*)
10.0	Para grass (*Brachiaria mutica*)
9.6	Blue panic (*Panicum antidotale*)
9.2	Hybrid Napier (*Pennisetum purpureum*), Setaria grass (*Setaria anceps*)
8.8	Anjan grass (*Cenchrus ciliaris*), Guinea grass (*Panicum maximum*) Deenanath grass (*Pennisetum pedicellatum*)
Forage crops (Rainy season, *kharif*)	
9.6	Sorghum (*Sorghum bicolor*), Sudan grass (*Sorghum sudanence*), *Dhaincha* (*Sesbania aculeate*)
9.2	Teosinte (*Euchlaena americanum*)
8.8	Cowpea (*Vigna unguiculata*), Guar (*Cyamopsis tetragonalob*)
Forage crops (Winter season, *rabi*)	
9.6	Oat (*Avena sativa*), Barley (*Hordeum vulgare*), *Shaftal*/Persian clover (*Trifolium resupinatum*)
9.2	*Berseem* (*Trifolium alexandrinum*)

The pH values at which yield reductions is 50% or less

Table 5.13 Relative tolerance of tree species to soil alkalinity/pH

Average pH_2 (0-120 cm)	Fuel-wood/Timber species#	Fruit trees
More than 10.0	*Prosopis juliflora* (*Pahari kiker*) *Acacia nilotica* (*Kikar*), *Casuarina equisetifolia* (Australian pine), *Tamarix articulata* (*Phrans*) *Casurina equietifolia* (*Jor-tor*)*	-
9.5 to 10.0	*Terminalia arjuna* (Arjan) Pithecello biumdulce (*Jangli jalebi*) *Eucalyptus tereticornis* (*Safeda*) *Albizzia lebbek* (*Siris*) *Tamarindus indica* (*Imly*) *Acacia catechu* (*Khair*) *Butea monosperma* (*Dhak*) *Cordia rothii* (*Lasura*) *Pongamia pinnata* (*Papri, Karanj*) *Parkinsonai aculeata*	*Achras japota* (sapota) *Carissa carandus* (*Karaunda*) *Psidium guajava* (Guava) *Emblica officinalis*(Aonla) *Zizyphus maurtiana* (*Ber*) *Syzgium cumini(Jamun)* *Phoenix dectylifera* (*Khajoor*) *Aegle marmelos* (*Bael*)
9.0 – 9.5	*Dalbergia sisoo* (*Shisham*) *Azadirachta indica* (*Neem*) *Grevillia robusta* (Silver Oak) *Sesbania sesban* (*Dhaincha*) *Kigelia pinnata* (*Kigelia*) *Melia azedarach* (*Bakain*) *Leucaena leucocephala* (*Subabul*) *Moringa olifera* (*Saunjana*) *Bambusa spp.* (Bamboo) *Morus alba* (Mulberry) *Albizia procera (Siris)* *Acacia auriculiform, Acacia senegal*	*Punica granatum* (Pomegranate) *Prunus persica* (Peach) *Pyrus communis* (Pear) *Vitis vinifera* (Grapes) *Mangifera indica* (Mango)
< 9.0	*Tectona grandis* (Teak) *Populus deltoides* (*Poplar*) *Ficus virens, Bombax ceiba* *Hardwickia binnata, Thespesia populnea, Bauhinia variegate* *Lagerstroemia speciosa, Cadrella toona*	

*Not suitable for deep groundwater areas

Genotypic Differences

Genotypes differ widely in their tolerance to both salinity and alkalinity. Therefore, besides selection of crops, selection of crop varieties is equally important. Choice of the tolerant varieties within a crop can result in more production and better economic returns from salt affected soils. Some of the promising cultivars and tolerance of few of them to soil pH are shown in Table 5.14. A field view of the salt tolerant varieties of rice, wheat mustard and chickpea (*Chana*) is shown in Fig. 5.2. Elaborate studies have been made to identify factors that affect growth, ionic composition and yield of crops in alkali soils (Sharma and Gill, 1994, Sharma, 2003, Singh and Sharma, 2005, Singh et al., 2018). Specific information on several varieties of crops developed at CSSRI, Karnal is presented in Table 5.15. Some of the drought and salt tolerant varieties of sugarcane are collated and listed in Table 5.16 (Gupta and Gupta, 2019).

Table 5.14 Promising varieties of important crops released for cultivation in saline and alkali soils

Crop	pH/ESP	Salinity, EC (dS m^{-1})	Varieties
Rice	9.8-10.2	10.0	CSR10, CSR 13*
	9.4-9.8		CSR23*, CSR27*, Basmati CSR30*, CSR36*, CSR 46, CSR 49, CSR 52, CSR 56, CSR 60
Wheat	9.2-9.3	6.5	KRL 1-4, KRL 19, WH157, KRL 210, KRL 213, Raj 3077, KRL 283
	8.7-9.0	5.5	HD 2009, HD2285, HD2329, WH542, C306
Indian mustard	8.8-9.3	6.5	Pusa Bold, Varuna Kranti, CS52*, CS54*, CS56*, CS 58, CS60,
Barley	up to pH 9.3	11.0	CSB1, CSB2, CSB3, DL200, Ratna, BH97, DL348
Sugar beet	9.5-10.0	10.0	Ramonskay 06, Poly rava – E, Tribal, Meribo – resistapoly
Sugarcane	<9.0		CO453, CO1341,CO6801, CO62329 and CO1111
Gram	<9.0		Karnal Chana 1

Source: Modified from Misra (1996)

Fig. 5.2 Field view of salt tolerant rice (top left), wheat (top right), mustard (bottom left) and chickpea (bottom right)

Table 5.15 Salt tolerant Rice, Wheat, Mustard and Chick pea varieties developed at CSSRI, Karnal

Crop/ Variety	Year of release	Seed yield under stress (t ha^{-1})	Salinity level (EC_e, dS m^{-1})	Alkalinity level (pH)
Rice				
CSR 10	1989	3.0	11.0	10.2
CSR 13	1998	3.0	9.0	10.0
CSR 23	2004	4.0	10.0	9.9
CSR 27	1998	4.0	10.0	9.9
CSR 30	2001	2.0	7.0	9.5
Basmati CSR 30	2012	2.0	7.0	9.5
CSR 36	2005	4.0	11.0	10.0
CSR 43	2011	3.5	7.0	9.5
CSR 46	2018	3.0	7.0	10.2
CSR 49	2018	2.9	7.0	10.2
CSR 52	2018	2.8	7.0	9.8
CSR 56	2018	3.0	11.0	10.2
CSR 60	2018	3.0	11.0	10.2
Wheat				
KRL 1-4	1990	3.0	7.0	9.3
KRL 19	2000	3.0	7.3	9.3
KRL 210	2010	3.5	6.6	9.3
KRL 213	2010	3.3	6.4	9.2
KRL 283	2018	4.5	-	9.3
Indian Mustard				
CS 52	1997	1.5	6-8.5	9.3
CS 54	2005	1.6	6-9	9.3
CS 56	2008	1.6	6-9	9.3
CS 58	2017	2.0	10-12	9.4
CS 60	2018	1.8	10-12	9.3
Gram				
Karnal *Chana* 1	1998	1.5	6.0	9.0

Table 5.16 Salt and drought tolerant varieties of sugarcane developed in India

Varieties	Tolerance/resistance characters	Developed at
Co 95003, Co 93005, Co 97008, Co 85019, Co 99004, Co 2001-13	Salt tolerant	ICAR-SBI, Coimbatore
Co 8145, Co 88019, Co 94008, Co 99004, Co 2001-13, Co 2001–15, Co 0212, Co 06027	Salt stress and drought/ rain fed	ICAR-SBI, Coimbatore
Co 0238, Co 05011	Salt stress and drought	ICAR-SBIRC, Karnal
Co 395, Co 453, Co 87263, Co 975, Co 453, Co 1148	Water logging and salt stress	ICAR-SBI, Coimbatore
Bo 99, Bo 106, B 37172	Salt stress and drought/ rain fed	RAU, Pusa, Bihar
CoPant 97222, CoPant 90223	Water logging/rain fed	GBPUAT, Pantnagar
Co 290, Co 86249, Co 94008 and D 109	Multiple- abiotic stress	ICAR-SBI, Coimbatore
CoSe 96234	Multiple stress	UPCSR, Shajahanpur

Application of piecewise linear model to relatively rate the salt tolerance of varieties is illustrated through the comparison of 5 genotypes of rice (Table 5.17). The results on the basis of ESP_t reveal that CSR 3 (a natural selection from indica types) is the most tolerant to alkali with an ESP_t of 26.16 while MI-48 is the least tolerant with an ESP_t threshold of only 8.02 (Table 5.17). It may be noted that same conclusion remains valid if compared for ESP_{50}. Similar test with 6 genotypes of wheat indicated that Kharchia 65 was the most tolerant both in respect of ESP_t and ESP_{50}. The order of sodicity tolerance is Kharchia 65 >KRL 4-13 > KRL 3-23 > KRL 1-4 = HD 2009 > HD 4502. There are not much significant differences among the last four genotypes. The yield potential of Kharchia 65 being quite low, this variety is not recommended for general cultivation but is used as a parent to develop salt tolerant wheat varieties. A view of the salt tolerant genotypes Basmati CSR 30 (rice) and KRL 19 (wheat) is shown in Fig. 5.2. In a recent study it has been observed that the rice genotypes CSR 10-M27, Bulk 212, CSR 2K 232, Bulk 18, CSR 11-117, CSR10 M2-27, NDRK 11-2, CSR 11-121, CSR 11-143 and CSR 27-192 are most stable in salt affected soils. The genotype Bulk 212 was found to be the highest yielding variety in most situations across the salt stress environments. These rice genotypes can be used for commercial cultivation in salt affected soils and also used in the further breeding program to transfer the salt tolerance to high yield varieties (Krishnamurthy *et al.*, 2015). Krishnamurthy *et al.* (2017) identified three genotypes namely CSR-2K-219, CSR-2K-262, and CSR-2K-242, which are superior and stable amongst all the tested genotypes with higher mean yields. Genotype CSR-2K-262 is the only genotype that performed best under both saline and alkaline environments.

Table 5.17 Sodicity tolerance of rice genotypes

	CSR-1 Mutant	MI-48	*Basmati* 370	CSR-1	CSR-3
Threshold ESP_t	20.36	8.02	15.92	20.63	26.16
Slope	2.03	2.22	1.84	0.82	0.73
ESP_{50}	45.00	30.60	43.10	81.60	94.65
Mean yield (t ha^{-1})	1.49	1.62	1.32	1.49	1.64

Without doubt, salt tolerant varieties can boost the land reclamation efforts not only in India but globally. But a major difficulty with salt tolerant varieties is that the window in which these varieties yield more than high yielding varieties is quite small (Fig. 5.3). Once the soil is reclaimed, a salt tolerant variety may yield less than the conventional high yielding variety. Besides, any further deterioration in the soil condition of un-reclaimed soils may not allow the variety to give economic yield.

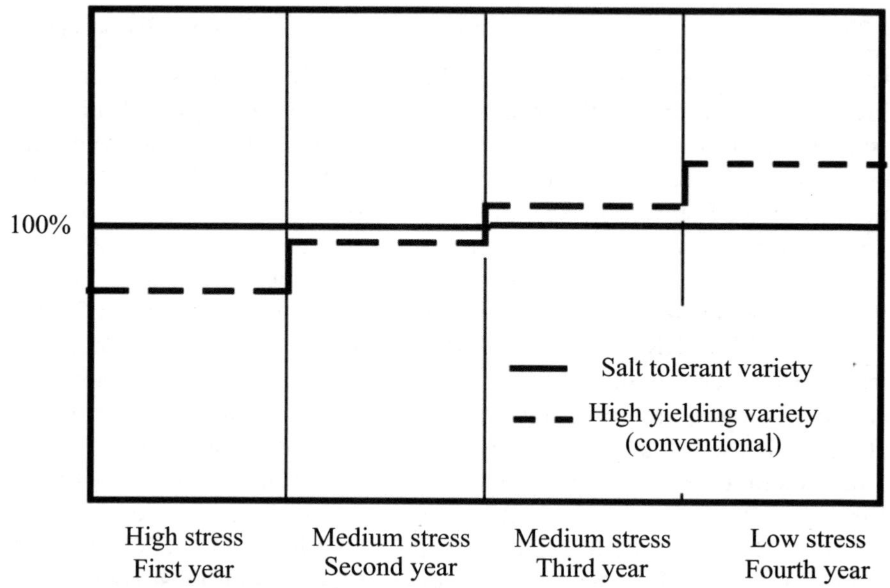

Fig. 5.3 Anticipated relative yields of a high yielding and a salt tolerant variety over years of reclamation in an alkali soil

In summary it may be stated that the salt tolerant varieties may not be treated as a substitute for good management of lands to prevent soil salinization or leaching. Moreover, information on salt tolerant varieties should be synthesized with reference to the local conditions under which these crops are cultivated. Since all the data on relative salt tolerance included in this chapter pre-supposes that normal management practices are employed, any major variation from one area to another may impact the varietal performance.

PHYTOREMEDIATION

Several investigators have taken forward the concept of phytoremediation through salt tolerant grasses (Kumar and Abrol, 1986) and forest plantations to reclaim alkali soils (Singh and Singh, 1993, Singh and Dagar, 2005). Significant reduction in some of the chemical (pH, ESP and EC) and favourable changes in the physical properties (Reduced bulk density and dispersion coefficients and increased infiltration rate) upon cultivation of the grasses and plants have been reported. Karnal grass, Rhoades grass, Bermuda grass and para grass have been reported to be promising grasses for large scale utilization of alkali lands, being able to grow at pH 10.4 except para grass which can be grown at a pH of 10.0. The studies further revealed that Karnal and para grasses have the great potential to phytoremediate an alkali soil for the following reasons:

- These grasses can be cultivated in highly alkali lands without application of amendments
- The grasses are tolerant to water logging (surface stagnation), which is a common phenomenon in alkali lands
- The lands are reclaimed within 2-3 years and thereafter cropping of rice-wheat crop rotation is possible beginning with salt tolerant rice and wheat varieties.

Moreover, the grasses grown on alkali soils have been found to be palatable to animals, have high yield and can be used as fodder.

For the forest plantations, technology of spot reclamation of alkali soils has been perfected. Any of the 3 methods namely auger hole method, pit method or a combination of pit and auger hole methods can be used. Auger hole method is easier to apply. Besides, providing fuel wood and biomass, growing of trees helps to ameliorate alkali soil. It is attributed to fall of leaf litter and other foliage, which upon decomposition brings in favourable changes in the physico-chemical properties of the soils. Apparently, the degree to which a salt affected soil is ameliorated by afforestation and agro-forestry practices depends upon the tree species, nature of growth, quality and quantity of litter, ability of tree species to fix nitrogen, planting density, age of plantation and other management practices. Trees species such as *Prosopis juliflora*, *Acacia nilotica*, *Prosopis chinensis*, *Acacia auriculiformis*, *Azadirachta indica*, *Casurina equisetifoia*, *Dalbergia sissoo*, *Alianthus excelsa*, *Prosopis cineraria*, and *Acacia tortilis* can be used for the phytoremediation of the alkali soils. Some of the pertinent observations are:

- The reclamation is faster and greater under *Prosopis juliflora* compared to other tree species
- Growing leguminous trees like *Prosopis, Acacia* and *Casuarina* ameliorates these soils at a much faster rate than non-leguminous trees, because of the build-up of organic matter and the recycling of important nutrients.

It has been established that highly alkali soils reclaimed through tree plantations can be successfully used for growing all kinds of agricultural crops after one cycle of plantation.

ENHANCING PLANTS' ADAPTATION MECHANISMS

Excessive salt concentration affects the plants either as a result of general osmotic stress or due to specific ion stress notably the excess Na and Cl. Plants respond to stress either through stress avoidance or stress tolerance through acclimatization and adaptation mechanisms. One of the ways to overcome the osmotic stress is through osmotic balance in the cytoplasm maintained by the accumulation of organic solutes (compatible osmolytes) such as sugars (sucrose, fructose, trehalose), ions (K^+), charged metabolites (glycinebetaine) and amino acids (proline) which do not inhibit metabolic processes. Besides, plants produce low molecular weight antioxidants such as ascorbic acid, reduced glutathione, tocopherol, carotenoids and detoxifying enzymes to scavenge reactive oxygen species (ROS). The studies are underway to unravel the complex physiological traits, metabolic pathways, and molecular or gene networks leading to better understanding of biochemical, physiological and molecular mechanisms regulating plant's adaptation to salt stress to improve salinity tolerance in crop plants. These studies are precursor to the studies being undertaken on the exogenous application of some osmolytes (Proline, Glycinebetaine, mannitol, ascorbic acid or thiamine or sodium salicylate) to perform osmo-protective roles and has been suggested as an alternative approach to improve crop productivity under abiotic stress conditions (Ashraf and Foolad, 2007). Deivanai et al. (2011) demonstrated that rice seedlings from the seeds pretreated with proline exhibited higher salt tolerance. The proteins accumulated under salt stress conditions provide a storage form of nitrogen that is reutilized in post-stress recovery and also play a role in osmotic adjustments (Kumar et al., 2015). An improvement in growth after applying exogenous osmolytes has been observed in tobacco, wheat, barley, rice, soybean and common beans (Borojevic et al., 1980; Agboma et al., 1997; Lutts, 2000). Different proteomics and transcriptomics studies have also indicated the major contribution of antioxidant defense related genes or proteins for salt tolerance. However, the results are not uniform in all cases so that sometimes some antioxidant enzymes are related, while others are not related to salt tolerance. There is need to carry forward these studies to arrive at doable technologies to impart salt tolerance and hasten the varietal development programs (Zhawar and Kaur, 2018).

In an earlier section of this chapter, relative salt tolerance of crops has been described with the help of four criteria. Attempts have also been made to evolve a criterion of relative salt tolerance based on metabolic parameters. The sodium and potassium contents in plant tissues and particularly their ratio have been found to be suitable parameters to adjudge the relative tolerance especially to alkali conditions. Studies on few varieties known for their differential affinity to alkali conditions revealed that high K/Na ratio at tillering stage is an indicator of a tolerant variety compared to a variety which has low K/Na ratio (Joshi *et al.*, 1980). Kumar et al. (2017) reported that the varieties CSR 10 and CSR 36 (both salt tolerant varieties) accumulated less Na^+ ions in leaves under stress. The highest value of K^+/Na^+ ratio was recorded in CSR 36, whereas, it was lowest in IR 29 (a salt sensitive variety). Free proline status of leaves at seedling stage has also been found a good indicator of alkali tolerance (Bal, 1976). But in later tests, it turned out that this is not very much reliable. The change in pH of the cell sap as a result of alkali stress also has been found to be useful to compare relative/tolerance of crops and varieties (Dwivedi, 1979). The alkali halophytes showed only a nominal rise in pH of their cell sap at ESP 40 as compared to ESP 10. The tolerant *Kharchia* variety of wheat exhibited a rise of only 1% in its sap pH at ESP 40 at 35 days growth stage. In spite of the concerted efforts made in this direction, except for K/Na ratio, no other criteria have uniformly performed well. More information needs to be generated on such tests so as to arrive at a sound criterion.

Chapter - 6

CHARACTERISTICS AND MANAGEMENT OF POOR QUALITY WATERS

INTRODUCTION

Irrigation is an essential element of high yield sustainable agriculture systems in India. Government of India therefore invested heavily on expansion of irrigation. These efforts paid rich dividends albeit with few areas of concern. Besides, anticipated inadequacy of water is likely to set in motion intense competition for the fresh water amongst its users. It is for sure that agriculture, which is unable to pay the economic cost of water, will have to forego its share of water in favour of other users. At the same time increasing water requirement of agriculture is likely to cause a huge gap between the supply and demand within the agriculture sector. Apparently, agricultural sector has to look for alternative sources of water. One of the alternative sources is the so called poor quality waters especially the naturally occurring saline and alkali groundwater commonly encountered in arid, semi-arid and coastal regions.

Quality rating of any water depends upon its intended use. For agriculture, water is categorized as good, if its long-term use sustains crop productivity and it has no adverse impact on the soil resource or product quality. Systematic studies in India have proved beyond doubt that waters categorized as unusable as per conservative guidelines can be effectively used for irrigation in monsoon climatic conditions of India. This chapter describes the characteristics of various kinds of water encountered in nature. Scientifically proven management practices are collated and included for effective use of these waters for irrigation.

WATER SOURCES AND THEIR QUALITY

The rainwater is the purest form of water, having the lowest salt content of all types of water used for irrigation. However, it may contain some dissolved gases and dissolved salts originating from terrestrial and marine sources. Generally, the amount of ions in rainwater varies depending upon the distance from the sea. With increasing industrialization, concerns are being expressed on increasing incidences of acid rains. Many reports have emerged from around the globe documenting the adverse effects of acid rains on crops.

Just opposite is the sea water, a complex saline solution containing a large number of elements (ions, gases, organic matter, micro-fauna and flora, etc.). Out of 72 elements so far detected, only nine are present in sufficient quantities in the sea water. The geo-chemical nature of sea water is Na-Mg-Cl-SO_4 type.

The river water is normally fresh water. The salt content and kind of salts present in the river water are governed by the type of rocks it encounters during its journey, nature of the soil over which the water flows and of course the climatic conditions. However, it may be occasionally polluted due to external factors. In general, canal waters that originate from the rivers or their reservoirs have similar quality that represents the parent Rivers, unless contaminated otherwise. Therefore, canal water is as fresh as river water.

The drains, either natural or manmade, carry excess rainwater during the monsoon season. Quality of the drain water may be as fresh as the canal water. However, drainage water emanating from irrigation runoff in few southern canals and drainage water in few channels in Uttar Pradesh show that drainage water can be moderate to high in salinity and high in SAR/RSC. During post monsoon period, most of drains in India only carry sewage water or industrial effluents generated from the adjoining urban centers and industrial units. The quality of this water needs to be examined before its use for irrigation.

Table 6.1 Estimated area (M ha) underlain with saline groundwater (EC >4 dS m^{-1})

State	Total area of the state (M ha)	Area underlain with saline groundwater (M ha)
Rajasthan	34.2	14.10 (41.22)
Gujarat	19.6	2.43 (12.39)
Haryana	4.4	1.14 (25.90)
Karnataka	19.1	0.88 (4.60)
Tamil Nadu	13.0	0.33 (2.53)
Punjab	5.0	0.30 (6.00)
Uttar Pradesh	2.9	0.13 (4.48)
Delhi	0.14	0.01 (7.14)
Total (Investigated, India)	125.0	19.3 (15.44)

Note: Figures in parenthesis are in percentage of the total area of the state. Source: Sharma (2000)

The groundwater quality has been found to vary both at the macro and micro scale showing large variations even within short distances. Nonetheless, major fraction of groundwater is fresh in nature, its EC varying in the range of 0.5-2.0 dS m^{-1}. Extent of saline groundwater in few states provides a glimpse into the problem of saline water in these states of India (Table 6.1). The degree and kind of mineralization of the groundwater is governed by the diverse hydrometeorology, topographical, drainage, geohydrology conditions as well as from activities like surface water irrigation and overexploitation of coastal aquifers. The marginal and poor quality waters constitute a greater part of phreatic groundwater resources in arid, semi-arid and coastal regions. In some parts of western Rajasthan, highly saline groundwater occurs naturally, so much so that in some cases chloride content may exceed 30 meq L^{-1}. The groundwater in the coastal territories of Saurashtra and Kutch is highly saline because of the sea water intrusion occurring as a result of over exploitation of groundwater. The distribution of salty water in India categorized as saline, high SAR saline and alkali water is shown in Fig. 6.1.

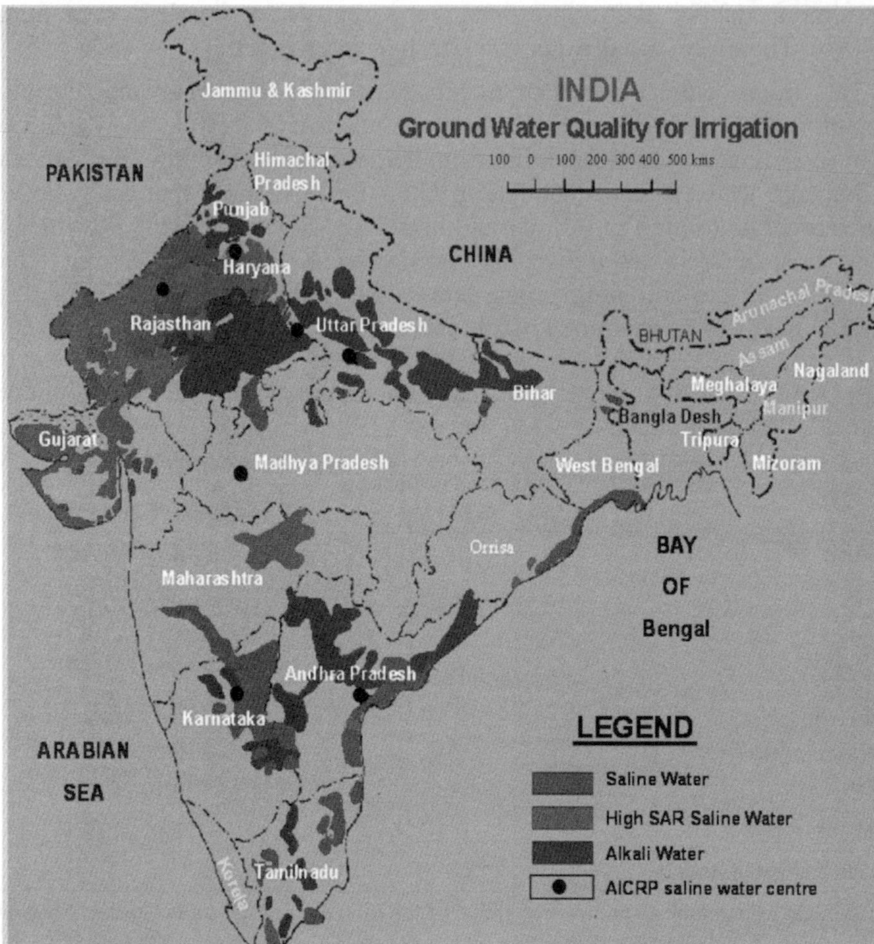

Fig. 6.1 Groundwater quality map for irrigation (First Approximation)

WATER CHARACTERISTICS AND ASSOCIATED PROBLEMS

The water characteristics/quality parameters such as EC, pH, SAR, RSC and toxic elements are useful to classify waters in different quality groups for irrigation purpose. Suitability of these groups of waters within the framework of irrigation water quality guidelines is decided on the basis of above mentioned quality parameters taking into account climate conditions, irrigation practices, soil water retention characteristics, crop tolerance, depth to water table, agronomic practices, etc. Laboratory procedures to determine some of the parameters to assess the problems related to salinity, sodicity and alkalinity are included in Chapter 7. Following sections deals with the salinity, sodicity and alkalinity hazards besides the permeability problems that may be encountered from continuous use of various qualities of water for irrigation.

Salinity Hazard

The total concentration of soluble salts is the single most important criterion which is conventionally used to determine the quality of irrigation water. Quantitatively, it is measured in terms of EC, because it is very closely related with the sum of major cations (or anions) determined by chemical analysis, and because this correlates well with the value of total dissolved solids, as well as osmotic potential. It is customary to denote irrigation water salinity as EC_{iw}. Based on EC_{iw}, waters on the basis of their utility and source are categorized as listed in Table 6.2.

Table 6.2 Classification of waters based on EC_{iw}

Water class	EC_{iw} (dS m^{-1})	Type of water
Non saline	<0.7	Drinking/irrigation water
Slightly saline	0.7-2.0	Irrigation water
Moderately saline	2.0-10.0	Drainage/groundwater
Highly saline	10.0-25.0	Drainage/groundwater
Very highly saline	25.0-45.0	Very saline groundwater
Brine	>45.0	Sea water

Per se saline irrigation water is not harmful to the crops except in sprinkler irrigation. On the other hand, when saline water is continuously used for irrigation, there is build-up of salts in the root zone in proportion to the EC_{iw} that becomes detrimental to crop growth. For example when 5 cm of irrigation water having the EC of 1 dS m^{-1} is applied to a soil; it adds about 320 kg of the salts as follows:

$$0.64 \text{ kg m}^{-3}(1\times 640 \text{ mg L}^{-1}) \times 100\times 100\times 5/100 = 0.64\times 500 = 320 \text{ kg ha}^{-1}$$

Thus, salt load applied per ha during the cultivation of wheat (400 mm of irrigation water) will be 2,560 kg and during cultivation of rice (800 mm of irrigation water) will be 5120 kg with a total of 7680 kg for the rice-wheat rotation.

If the irrigation water applied remains within the root zone i.e. there is no leaching, then the salinity levels will continue to rise linearly with each irrigation. In this case, change in EC_e at any time will mainly be governed by the EC_{iw} besides the depth of water applied and depth of the root zone. The change in EC_e can be calculated with the following equation:

$$\Delta EC_e = D_i/D_s \times \rho_w/\rho_b \times EC_i \times 100/SP \tag{6.1}$$

Here ΔEC_e is the change in salt concentration of the soil layer, D_s, cm, D_i is the amount of irrigation water applied, cm, SP is the water content in mass percentage at saturation, ρ_b and ρ_w are the bulk density of soil, g cm^{-3}, and density of water respectively which is usually taken as 1g cm^{-3}. In general, irrigation water having EC of 2 dS m^{-1} can be successfully employed for crop cultivation on sustainable basis. Water having EC beyond this limit result in yield losses depending upon the kind of crop cultivated, kind of soils and the management practices adopted. General guidelines to assess the salinity hazard of water are given in Table 6.3.

Table 6.3 EC_{iw} and degree of salinity hazard

Total dissolved solids (TDS) (ppm)	EC_e (dS m^{-1})	Hazard
< 500	< 0.75	None
500-1000	0.75-1.5	Slight
1000-2000	1.5-3.00	Moderate
> 2000	> 3.0	Severe

Soil Salinization

All the salts applied along with irrigation water are not retained in the root zone as postulated in deriving eq. (6.1). Some of the water does move beyond the root zone as deep percolation loss. Thus, a fraction of the salts applied with irrigation water also go below the root zone along with the water. Whatsoever salts are retained, these salts over a period of time accumulate in the root zone and may adversely impact the crop growth. The soil's EC_e values irrigated with saline water have been observed to be 2 to 10 times more as compared to the EC_{iw}. Under steady state conditions, the relation between EC_e and EC_{iw} is given by the relation:

$$EC_e = A \times EC_{iw} \tag{6.2}$$

Here 'A' is a concentration factor, which depends upon the leaching fraction (LF). The relevant concentration factors as controlled by leaching fraction are reported in Table 6.4 (Ayers and Westcot, 1985). The data show that the steady state salinity can be about 3.2 times the salinity of the irrigation water at low leaching fractions while it can be even less than the salinity of the irrigation water at high leaching fraction. At a leaching fraction of 0.30, concentration factor is unity i.e. EC_e will be equal to EC_{iw}. The steady state root zone soil salinity (EC_e) likely to develop upon long-term use of saline water at various leaching fractions (LF) and for a water uptake pattern of 40:30:20:10 is shown in Fig. 6.2. The water uptake pattern 40:30:20:10 envisages that the total water used by the crop is derived from the root zone in the ratio of 40, 30, 20 and 10 from each 25% of the root zone beginning from the soil surface (Fig. 6.3). If the water uptake pattern is different than 40:30:20:10, then concentration factors need to be calculated. Procedure to calculate the concentration factors for various leaching fractions and for any water uptake pattern is illustrated in Example 6.1. For a water uptake pattern of 60:25:10:5, concentration factor will be slightly higher than with 40:30:20:10 pattern.

Table 6.4 Concentration factor (A) for use in eq. (6.2)

Leaching fraction (LF)	Applied water needed (% of ET)	Concentration factor (A)
0.05	105.3	3.2
0.10	111.1	2.1
0.15	117.6	1.6
0.20	125.0	1.3
0.25	133.3	1.2
0.30	142.9	1.0
0.40	166.7	0.9
0.50	200.0	0.8
0.60	250.0	0.7

Source : Ayers and Westcot (1985)

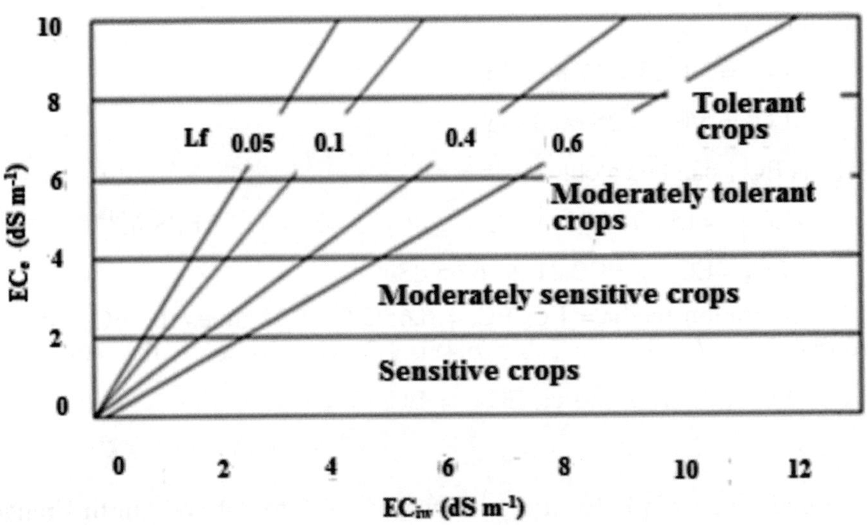

Fig. 6.2 Steady state root zone soil salinity (EC_e) likely to develop with applied irrigation water salinity (EC_{iw}) at various leaching fractions (LF)

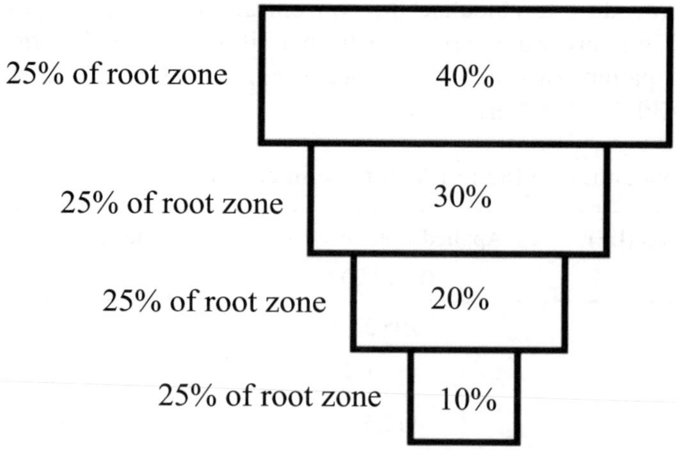

Fig. 6.3 Water uptake pattern (40:30:20:10) by the plants

Example 6.1

A crop is irrigated with a water of EC_{iw} 5.0 dS m^{-1}. If the LF is 0.2, calculate the average root zone salinity assuming the water uptake pattern of 40:30:20:10.

EC_1 (Fig. 6.3) = 4.0 /1.0 = 5.0/1.0 = 5.0 dS m^{-1} (LF = 100%)

EC_2 =5.0/(1-(0.40) (1-0.20)) = 5.0/0.68 = 7.35 dS m^{-1}

EC_3 = 5.0/ (1-(0.40+0.30) (1-0.20)) = 5.0/0.44 = 11.36 dS m^{-1}

Similarly

EC_4 = 5/0.28 = 17.86 dS m^{-1}

EC_5 = 5/0.2 = 25.00 dS m^{-1}

Thus EC_{sw} can be calculated as

EC_{sw} = (5.00+7.35+11.36+17.86+25.00)/5 =13.31 dS m^{-1}

or $EC_e = EC_{sw}/2$ = 13.31/2= 6.65 dS m^{-1}

Concentration factor = EC_e/EC_i = 6.65/5.0 =1.31 times the EC_{iw}. It may be noted that the value given in Table (6.4) is 1.3, which is nearly the same.

EC of drainage water, EC_{dw} = EC_i/LF = = 25.0 dS m^{-1}

pH

The normal range of pH for irrigation water is 6.5 to 8.4. Within this range the crops do well. Irrigation waters having pH outside this range may still be satisfactory but other problems of nutrition or toxicity become suspect. The pH therefore, is seldom of any real importance by itself. The main use of pH is in quick evaluation of the possibility that the water may be abnormal. If an abnormal value is found, this should be a warning that the water needs further evaluation for possible corrective measure. For example, high pH above 8.5 is often associated with high bicarbonate (HCO_3^-) and carbonate (CO_3^{2-}) concentrations, which might turn the water alkali in nature. On the other hand, role of pH for pressurized irrigation system is quite important. Clogging of emitters increases with increasing pH. To minimize clogging problems it must be brought to 7+, if high.

Sodium Adsorption Ratio (SAR)

Water might be suitable for irrigation on the basis of its EC but may not be suitable if sodium predominates. Thus, an important consideration is whether the use of irrigation water of given quality will cause sodicity hazard in the soil. It is usually assessed through SAR of the irrigation water denoted as SAR_{iw}. Water with high SAR upon its continuous use causes soil permeability problems. The infiltration rate may be reduced to an extent that the crop is not adequately supplied with water that may adversely impact crop growth (Gupta, 2015). The degree of adverse effect depends upon total electrolyte concentration, bicarbonate, carbonate and silica contents of irrigation water and clay mineral of the soil. Adverse effects on permeability will be much less if EC_{iw} is more and the bicarbonate, carbonate and silica content in the water are less. For stable permeability, the irrigation waters having SAR 10, 20, 30 should have total electrolyte concentration of 10, 20 and 30 meq L^{-1} (EC_{iw} 1, 2, and 3 dS m^{-1}) respectively. On the other hand, adverse effects will be more pronounced in soils having montmorillonite and illite-vermiculite (2:1 crystal lattice) minerals than soils having kaolinite-sesquioxides (1:1 crystal lattice).

From irrigation point of view, SAR_{iw} less than 15 $(mmol\ L^{-1})^{1/2}$ is normally considered safe. Recent reports have revealed that SAR_{iw} of even 10 might create problems in heavy textured soils. Therefore, some classifications of water on the basis of SAR_{iw} treat such water of medium hazard (Table 6.5). Although the SAR_{iw} concept has worked well and is widely used for its simplicity, the limitations of SAR_{iw} must be kept in view while making use of this parameter in assessing the sodium hazard.

Table 6.5 Classification of water based on SAR_{iw} values

SAR	Sodium hazard of water	Remarks
1-9	Low	Cautious use on sodium sensitive crops
10-17	Medium	Amendments (such as gypsum) and leaching needed
18-25	High	Generally unsuitable for continuous use
>26	Very high	Generally unsuitable for use

A major limitation of SAR concept has been discussed in Chapter 2, which questions the lumping together of Ca^{++} and Mg^{++} in calculating SAR. In case of SAR_{iw}, the concept may work well when Ca and Mg ions in water are nearly equal. In the vast majority of irrigation waters in the USA, where SAR_{iw} worked well to assess the sodium hazard, Mg has been only a minor constituent. The examination of chemical composition of the these waters revealed that Mg/Ca ratio was generally less than one and mostly it varied from 0.2 to 0.5. In India, groundwater used for irrigation may have magnesium 2-5 times more than the calcium. This fact may be taken into consideration as SAR_{iw} calculated for such waters may underestimate the sodium hazard of waters.

Attempts to correlate SAR_{iw} with SAR_e, the SAR of the soil saturation extract, have not met with success mainly because of the two reasons. One is the increase in soil water concentration over that of the applied irrigation water by evapotranspiration. Although, it does not change the relative proportions of sodium and calcium + magnesium ions, but will change the SAR_e. It is because of the square root relationship between sodium and calcium plus magnesium ions. SAR_e is related with SAR_{iw} through LF according to the following relationship:

$$SAR_e = SAR_{iw}/\sqrt{LF} = SAR_{iw}\sqrt{(EC_e/EC_{iw})} \tag{6.3}$$

Apparently, higher ESP values are likely to develop in soils than predicted using SAR_{iw}.

Rhoades (1972) pointed out that the effect of soil water composition on salt precipitation or mineral weathering is difficult to assess. The loss of calcium and magnesium due to precipitation as carbonates is a major factor affecting the final SAR_e value. Weathering of certain soil minerals may also affect the SAR of the soil water by introducing additional amounts of calcium, magnesium and bicarbonate. Besides, at similar concentrations, adverse effect due to SAR_{iw} is highest when the irrigation water contains Na_2CO_3 followed by $NaHCO_3$, Na_2SO_4 and NaCl. It is why the attempts to predict sodium adsorption ratio of drainage water (SAR_{dw}) below the root zone, which determines the adverse effects of SAR_{iw} on permeability problems, have not succeeded for their universal application.

Application of chemical amendments is recommended to mitigate infiltration problem caused by irrigation water having low EC_{iw} and high SAR_{iw}. Since the adverse effects primarily occur in the upper few cm of soil, it may be useful to apply some amendment to the soil on seasonal or annual basis. To be effective, the amendment should be mixed in a shallow soil depth.

Alkalinity Hazard

Residual Sodium Carbonate (RSC)

Eaton (1950) proposed the concept of residual sodium carbonate (RSC). He observed that when water containing CO_3+HCO_3 higher than Ca+Mg is used for irrigation, it may lead to the development of alkali soils (saline or non-saline) especially if water is used so sparingly that little leaching occurs. It is because there is a tendency for calcium and to some extent magnesium to precipitate as carbonates as the soil solution concentrates upon evapotranspiration. On the assumption that their precipitation is complete, a concept of residual sodium carbonate (RSC) has been proposed to evaluate the quality of bicarbonate containing water. RSC has been defined as:

$$RSC = (CO^-_3 + HCO^-_3) - (Ca^{++} + Mg^{++}) \tag{6.4}$$

All ions are expressed in meq L^{-1}, resulting in units of RSC as meq L^{-1}. Depending upon the relative proportions of Na and Ca in the water following two possibilities may be encountered.

If sodium is not too greatly in excess of Ca+Mg and enough of such water is used such that water contains a substantial proportion of Ca and Mg, there may be little ill effect as revealed by the following reaction.

$$Ca^{++} + 2HCO^-_3 = CaCO_3 + H_2O + CO_2 \uparrow \tag{6.5}$$

This reaction does not go to completion under ordinary circumstances, but in so far as it does proceed, the concentrations of calcium and magnesium are reduced with increase in the relative proportion of sodium. It may result in decrease in soil salinity and increase in ESP albeit slowly.

If sufficient sodium is present, the residual bicarbonate reacts with sodium and forms sodium bicarbonate which upon evaporation is converted into sodium carbonate according to the following reactions:

$$Na^+ + HCO_3 = NaHCO_3 \tag{6.6}$$

$$2NaHCO_3 = Na_2CO_3 + H_2O + CO_2 \uparrow \tag{6.7}$$

The sodium carbonate, thus formed reacts with normal soil and replaces exchangeable calcium with sodium on exchange complex (EX).

$$Ca^{++}(Ex) + Na_2CO_3 = 2Na^+ (Ex) + CaCO_3 \tag{6.8}$$

Formation of indurated layers of $CaCO_3$ has often been associated with the use of irrigation waters containing bicarbonate ions more than calcium ions. Over time, layers of carbonate deposits become indurated or hardened. These are essentially impermeable, except for cracks and other discontinuities. Indurated layers normally are a product of hundreds or even thousands of years of accumulation, so their formation during short-term agricultural operations is probably unimportant.

Richard (1954) recommended that both RSC and SAR should be used to evaluate the sodicity hazard. Similarly, Ayers and Westcot (1976) included RSC as a diagnostic criterion but their later publication is silent on this parameter (Ayers and Westcot, 1985). Subsequent investigators showed that general reliability of the RSC concept for evaluating the sodium hazard of bicarbonate water is questionable. They argued that this parameter only indicates whether the calcite is likely to be formed or not and has no diagnostic value in itself. It is ultimately SAR or more importantly Adj. R_{Na} that will govern the quality of irrigation water. Besides, question has also been raised on the way RSC is calculated. Since conditions for Mg to precipitate in soils do not prevail in agricultural lands, its inclusion in the calculation of RSC has been questioned.

Should Mg be taken out from the RSC equation is the one question that needs further investigation?

Recent investigations in India also support this view. It is shown that if the RSC exceeds 2.5 meq L^{-1}, Adj. R_{Na} comes into play to assess precipitation of calcium from the water and/or dissolution of native calcium of the soils. Nonetheless, RSC is still being widely used in India as a diagnostic parameter. Efforts are also being made to develop technologies to manage high RSC water. Past guidelines reported in the literature seem to be very low as compared to the evidences generated in India probably because of the monsoon climatic conditions. Most reports from India recommend that a water having RSC < 2.5 meq L^{-1} is usually safe to apply. The problems begin beyond this value and the degree of problem increases with increasing RSC (Table 6.6).

Table 6.6 Guidelines to assess RSC (meq L^{-1}) hazard of irrigation water

Locations	None	Slight to moderate	Severe
International	1.25	1.25-2.5	> 2.5
India	<2.5	2.5-4.0	>4.0

Adjusted SAR

To predict sodicity hazard for those water which contain appreciable amounts of Ca and HCO_3 having tendency to take out Ca as revealed by eq. (6.5), Ayers and Wescot (1975) proposed that sodicity hazard should be determined by Adjusted SAR_{iw} to be calculated as SAR_{iw} (1 + 8.4 – pH_c). The pH_c is the calculated pH of the irrigation water in contact with lime and in equilibrium with soil CO_2. The values of pH_c above 8.4 indicate a tendency to dissolve lime from the soil and values below 8.4 indicate a tendency to precipitate lime from the applied water. Later studies revealed that the proposed parameter failed to resolve some of the issues related to such waters. Therefore, use of this parameter has become obsolete (Ayers and Westcot, 1985).

Adjusted Residual Sodium

To assess the dissolution and precipitation of calcium, Suarez (1981) proposed an equation and designated it as adj. R_{Na}. Adj. R_{Na} is calculated by the following equation.

$$\text{Adj.}\,R_{Na} = \frac{Na}{[(Ca_x + Mg)/2]^{1/2}} \tag{6.9}$$

Where, Ca_x is the value of Ca in the water that takes care of EC_{iw}, HCO_3/Ca ratio and the estimated partial pressure of CO_2 in the surface few mm of the soil (Table 6.7). The Ca_x values represents meq L^{-1} of Ca expected to remain in the soil solution at equilibrium. It seems that certain modifications in this concept can make it a versatile tool to predict soil ESP under Indian monsoon conditions.

Table 6.7 Calcium concentration (Ca_x) expected to remain in near-surface soil-water following irrigation

Ratio of HCO_3/Ca	Salinity of applied water (EC_{iw}, dS m^{-1})											
	0.1	0.2	0.3	0.5	0.7	1.0	1.5	2.0	3.0	4.0	6.0	8.0
0.10	8.31	8.57	8.77	9.07	9.31	9.62	10.02	10.35	10.89	11.32	12.01	12.56
0.15	6.34	6.54	6.69	6.92	7.11	7.34	7.65	7.90	8.31	8.64	9.17	9.58
0.20	5.24	5.40	5.52	5.71	5.87	6.06	6.31	6.52	6.86	7.13	7.57	7.91
0.25	4.51	4.65	4.76	4.92	5.06	5.22	5.44	5.62	5.91	6.15	6.52	6.82
0.30	4.00	4.12	4.21	4.36	4.48	4.62	4.82	4.98	5.24	5.44	5.77	6.04
0.35	3.61	3.72	3.80	3.94	4.04	4.17	4.35	4.49	4.72	4.91	5.21	5.45
0.40	3.30	3.40	3.48	3.60	3.70	3.82	3.98	4.11	4.32	4.49	4.77	4.98
0.45	3.05	3.14	3.22	3.33	3.42	3.53	3.68	3.80	4.00	4.15	4.41	4.61
0.50	2.84	2.93	3.00	3.10	3.19	3.29	3.43	3.54	3.72	3.87	4.11	4.30
0.75	2.17	2.24	2.29	2.37	2.43	2.51	2.62	2.70	2.84	2.95	3.14	3.28
1.00	1.79	1.85	1.89	1.96	2.01	2.09	2.16	2.23	2.35	2.44	2.59	2.71
1.25	1.54	1.59	1.63	1.68	1.73	1.78	1.86	1.92	2.02	2.10	2.23	2.33
1.50	1.37	1.41	1.44	1.49	1.53	1.58	1.65	1.70	1.79	1.86	1.97	2.07
1.75	1.23	1.27	1.30	1.35	1.38	1.43	1.49	1.54	1.62	1.68	1.78	1.86
2.00	1.13	1.16	1.19	1.23	1.26	1.31	1.36	1.40	1.48	1.54	1.63	1.70
2.25	1.04	1.08	1.10	1.14	1.17	1.21	1.26	1.30	1.37	1.42	1.51	1.58
2.50	0.97	1.00	1.02	1.06	1.09	1.12	1.17	1.21	1.27	1.32	1.40	1.47
3.00	0.85	0.89	0.91	0.94	0.96	1.00	1.04	1.07	1.13	1.17	1.24	1.30
3.50	0.78	0.80	0.82	0.85	0.87	0.90	0.94	0.97	1.02	1.06	1.12	1.17
4.00	0.71	0.73	0.75	0.78	0.80	0.82	0.86	0.88	0.93	0.97	1.03	1.07
4.50	0.66	0.68	0.69	0.72	0.74	0.76	0.79	0.82	0.86	0.90	0.95	0.99
5.00	0.61	0.63	0.65	0.67	0.69	0.71	0.74	0.76	0.80	0.83	0.88	0.93
7.00	0.49	0.50	0.52	0.53	0.55	0.57	0.59	0.61	0.64	0.67	0.71	0.74
10.00	0.39	0.40	0.41	0.42	0.43	0.45	0.47	0.48	0.51	0.53	0.56	0.58
20.00	0.24	0.25	0.26	0.26	0.27	0.28	0.29	0.30	0.32	0.33	0.35	0.37
30.00	0.18	0.19	0.20	0.20	0.21	0.21	0.22	0.23	0.24	0.25	0.27	0.28

Adapted from Suarez (1981); Assumes a soil source of calcium from lime ($CaCO_3$) or silicates; no precipitation of magnesium, and partial pressure of CO_2 near the soil surface (pCO_3) is 0.0007 atmosphere; Ca_x, is in meq L^{-1}.

Minhas and Sharma (2006) investigated the applicability of the concept of Adj. R_{Na} to predict the soil ESP developed as a result of medium to long-term use of high RSC water. They observed that the alkali hazard of these waters can be predicted using this concept. However, they suggested that for the monsoon climatic conditions of India, some dilution and dissolution processes commonly occurring under field conditions should be accounted for. They modified eq. (6.5) by considering a concentration factor, CF (due to varying ET) and a simple dissolution factor, DF (due to rainfall) and arrived at the following equation:

$$\text{Adj. } R_{Na}\text{ (mod.)} = [(\text{Diw/Drw})\ \{\text{ET/(Drw+ET)}\}^{1/2}] \times \text{Adj. } R_{Na} \qquad (6.10)$$

or

$$\text{Adj. } R_{Na}\text{ (mod.)} = A \times \text{Adj. } R_{Na} \qquad (6.11)$$

The values of A for the data reported by Minhas and Sharma (2006) for various cropping systems and climatic conditions are given in Table 6.8. The values besides crop rotation vary with rainfall, irrigation water applied and ET. It may be noted that when the rainfall is in 60-70 cm range, factor A follows the order Rice-wheat > Maize/millet-wheat > cotton-wheat >fallow-wheat. If rainfall is less (30-40 cm) as in two cases of fallow-wheat, the factor in fallow -wheat is even higher than maize/millet-wheat. In all cases a one to one relation between the soil ESP and Adj. R_{Na} (mod.) was observed. Since all data sets reported in Table 6.8 pertains to alluvial soils of Punjab and Haryana, factors for other soils and climatic conditions need to be worked out.

Table 6.8 Values of A of eq. (6.11) for various agro-climatic and cropping conditions

Rainfall (D_{rw}) (cm)	Irrigation water (D_{iw}) (cm)	ET (cm)	Modification factor, $A = [(\text{Diw/Drw})\ \{\text{ET}/(D_{rw}+\text{ET})\}^{1/2}]$
71	38	40	0.89 (Fallow-wheat)
70	38	40	0.90 (Fallow-wheat)
35	40	42	1.55 (Fallow-wheat)
30	40	42	1.75 (Fallow-wheat)
72	53	105	0.96 (Cotton-wheat)
60	130	105	2.72 (Rice-wheat)
72	137	100	2.50 (Rice-wheat)
77	142	100	2.45 (Rice-wheat)
70	56	80	1.10 (Maize/millet-wheat)
72	62	80	1.19 (Maize/millet-wheat)

PERMEABILITY PROBLEMS

Alkali water having high RSC (> 2.5 meq L^{-1}) used continuously over time increase soil ESP and pH and adversely impacts the infiltration rate of the soils. Irrigation to cotton-wheat in Punjab with water having EC 1.4 dS m^{-1}, SAR 14.9, RSC 10.1 meq L^{-1} over a period of 12 years increased the soil ESP from less than 5 to around 30 with application of 62 irrigations using 3720 mm of water (Fig. 1.3). Similar increase in ESP has been observed at Hisar. Irrigation water applied on such fields keeps on standing over the land for days together (Fig. 6.4).

Fig. 6.4 Water stagnation following an irrigation event in alkali water irrigated plot at Hisar

A first guess of the permcability problems associated with irrigation water can be assessed using any one of the following four diagrams (Fig. 6.5 to Fig. 6.8).

Rhoades Diagram

Swelling and dispersion occurring as a result of high exchangeable sodium developing on account of high SAR water is counteracted by high electrolyte concentration. Thus, the soil permeability hazard cannot be assessed independently of the electrolyte concentration of the irrigation water. Quirk and Schofield (1955) reported that for stable permeability of the soil, it will be necessary to maintain some threshold electrolyte concentration (EC_{iw}) in the irrigation water. This threshold electrolyte concentration increases with increasing SAR or soil ESP. Rhoades (1983) pointed out that infiltration rate being the property of the soil surface, SAR of the surface layer will determine likelihood of any reduction in the infiltration rate viz-a-viz the electrolyte concentration of the irrigation water. He devised a diagram that is commonly used to assess the permeability hazard (Fig. 6.5). In this diagram, we have a threshold line. Permeability problem is likely to occur if a particular SAR-EC combination of water lies to the left of the threshold line. No problem is likely to occur if the combination falls to the right of this line. Since SAR and EC_e of the top soil is normally in near equilibrium with SAR and EC_{iw} of the irrigation water, it is concluded that the irrigation waters having SAR, 10, 20 and 30 should not have TCC less than 10, 20 and 30 meq L^{-1} or EC not less than 1, 2 and 3 dS m^{-1}, respectively.

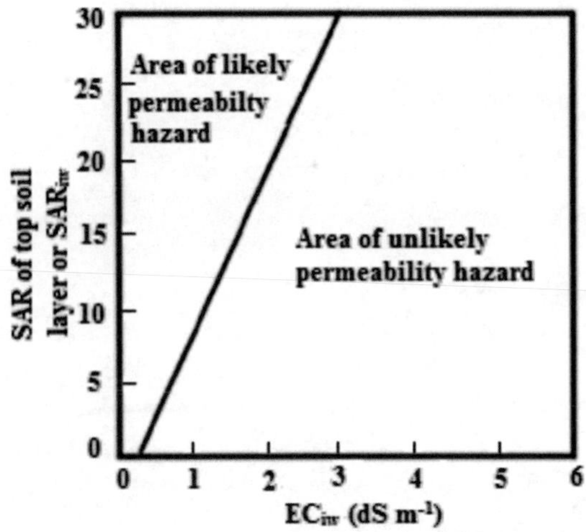

Fig. 6.5 Water infiltration hazard as assessed from EC_{iw} and SAR of irrigation water

FAO Guidelines

According to these guidelines, waters of various SAR-EC combinations have been grouped into 3 instead of 2 classes. The 3 classes have combinations of SAR-EC of irrigation water such that there is no reduction, slight to moderate reduction and severe reduction in infiltration rate. For almost no reduction, irrigation waters having SAR < 3, 3-6, 6-12, 12-20 and 20-40 should have EC greater than 0.7, 1.2, 1.9, 2.9 and 5.0 (dS m^{-1}), respectively (Fig. 6.6).

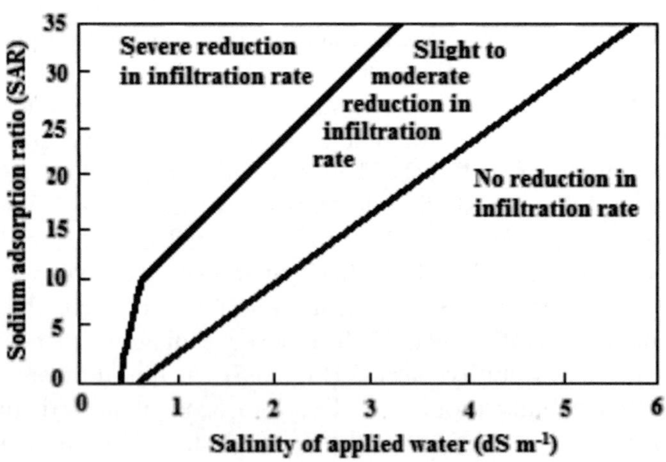

Fig. 6.6 Interrelation of SAR and EC of water to assess water infiltration problems

Soil Structural Stability

A diagram showing relationship between SAR and EC has been proposed to assess soil structure stability as a result of irrigation with poor quality water (Fig. 6.7). Water quality that falls to the right of the dashed line is unlikely to cause soil structural problems (ANZECC, 2000). Water quality that falls to the left of the solid line is likely to induce soil structure degradation. Appropriate corrective steps such as application of gypsum or some suitable amendment may be required. Water that falls between the lines is of marginal quality, and needs to treated with caution considering soil properties and rainfall.

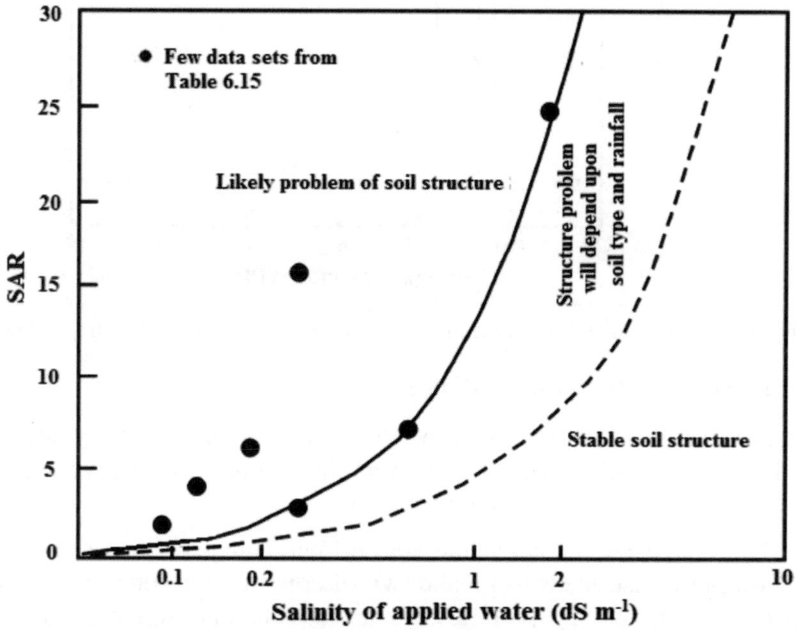

Fig. 6.7 Combinations of SAR and EC of irrigation water and soil structural stability

Permeability Index

According to Doneen (1963), the permeability Index (PI) of the water is calculated by the following equation:

$$PI = 100\ (Na + \sqrt{HCO_3}) / (Na + Ca + Mg + K) \quad (6.12)$$

It may be noted that unlike other equations/diagrams, it takes into account bicarbonate content of the water but it does not take into account the carbonates content of the water. Nonetheless, it may be possible to modify the equation by adding K and CO_3 in the numerator of eq. (6.12). To classify water on this basis, PI is plotted against total cation concentration. Water is then classified into 3 groups namely Class 1, which has no problem, class II with permeability reduction of about 25% and Class III with permeability reduction of about 75% (Fig. 6.8).

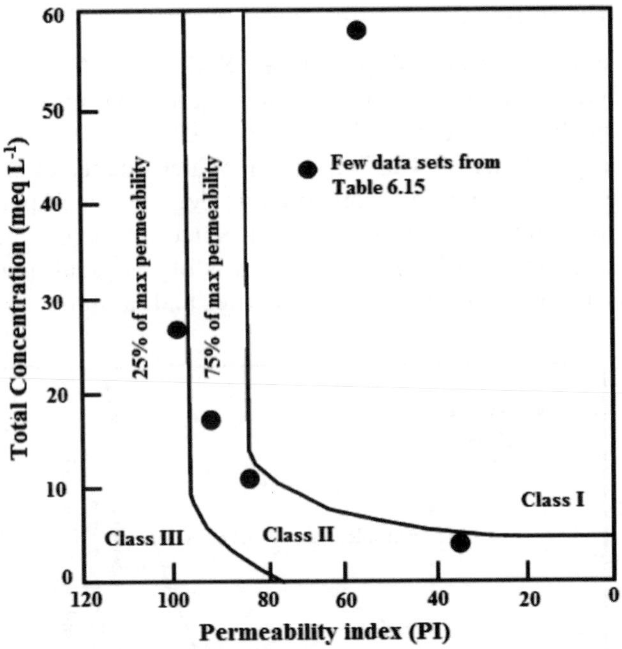

Fig. 6.8 Classification of irrigation water on the basis of permeability index

Harmful Effects of Alkali/Sodic Waters

Irrigation with alkali or sodic water is known to adversely affect crop growth. It is because of a number of factors that becomes operative mainly as a result of high ESP developing in the root zone.

- The increased exchangeable sodium percentages (ESP) resulting from long-term use leads to breakdown of soil structure due to swelling and dispersion of clay particles. Thus, less air and water enters the soil resulting in poor environment for root growth. Restricted internal drainage may cause temporary water logging. Such soils fail to easily attain proper soil moisture conditions for cultivation.
- A thin crust formed at the soil surface acts as a barrier to water penetration and seedling emergence.
- Crop growth may be affected as a result of direct toxic effects of sodium, carbonate and bicarbonates present in the water.
- The high ESP resulting from irrigation may cause imbalance in the uptake of various ions or plant nutrients.
- Increase in soil pH reduces the availability of a number of plant nutrients like nitrogen, zinc, iron, etc. It may also decrease availability of calcium, magnesium and potassium.

In most cases, the yield of crops is adversely affected by a combination of the factors described above.

TOXICITY HAZARD

The toxic waters have variable EC, SAR and RSC but have excess of an individual ion that causes toxicity to plants. A toxicity problem is different than the problems due to salinity or sodicity as the toxicity occurs within the crop as a result of the uptake and accumulation of toxic constituent of the irrigation water. The constituents of concern are: chloride, sulphate, sodium, nitrate, boron, fluoride, or heavy metals such as selenium, cadmium, lead and arsenic etc. High concentration of these ions may also cause imbalance in the uptake of various ions or even reduce the uptake of plant nutrients. Although the adverse effects may occur even at low salinity or sodicity but the toxicity problem accompanying salinity or sodicity complicates the issue with simultaneous development of salinity and infiltration problems. The toxicity effect depends on the type of crop, stage of growth, concentration of the toxic ion or ions, climate and soil conditions. Few constituents of water that may cause toxicity problems are briefly mentioned in the following paragraphs.

Chloride

Chloride is essential to plants albeit in low amounts. At high concentrations, it causes toxicity to sensitive crop. Classification of irrigation water on the basis of chloride content is reproduced in Table 6.9 (Maas, 1984).

Table 6.9 Classification of irrigation water on the basis of chloride content

Chloride (ppm)	Effect on crops
Below 70	Generally safe for all plants
70-140	Sensitive plants show injury
141-350	Moderately tolerant plants show injury
Above 350	Can cause severe problems

Bicarbonate

Bicarbonate water besides causing precipitation of calcium as calcium carbonate discussed in a previous section may have direct toxic effect on water if present beyond the permissible limit. Direct toxic effects of bicarbonate may appear only if bicarbonate in water exceeds 1.5 meq L^{-1}. Slight to moderate effects appear in the concentration range of 1.5 to 7.5 meq L^{-1} beyond which the toxicity effects may be severe.

Sodium

Primarily influence of Na^+ to cause toxicity is through inhibition of metabolic activity within the cytoplasm. Direct toxicity effect of sodium may appear if sodium concentration in water is more than 3 meq L^{-1}. Moderate to severe effects appear in the concentration range of 3-9 meq L^{-1} beyond which toxic effects may be severe.

Boron

Boron is another element that is essential in low amounts, but toxic at higher concentrations (Table 6.10). In fact, there is a very thin margin between the essentiality level and toxicity level of boron. Toxicity on sensitive crops, through its effect on the metabolic activities of the plants can occur in the range of 0.3 to 1.0 mg L^{-1}boron.

Table 6.10 Tolerance of selected crops to boron

Crop type	Sensitive		Moderately sensitive	Moderately tolerant	Tolerant
Maximum tolerable concentration (mg L^{-1})*	0.5-0.75	0.76-1.0	1.1-2.0	2.1-4.0	4.1-6.0
Crops	Peach	Wheat	Carrot	Lettuce	Alfalfa
	Onion	Barley	Potato	Cabbage	Sugar beet
		Sunflower	Cucumber	Corn	Tomato
		Dry bean		Oats	

Source: Maas (1987); *Maximum concentrations tolerated in soil water or saturation extract without yield or vegetative growth reductions.

Nitrogen

Nitrogen in irrigation water (N) is also largely a fertility issue, as nitrate-nitrogen (NO_3-N) can be a significant N source. Waters high in N result in excessive vegetative growth. It may also impair product quality in crops such as barley and sugar beets. However, these problems can be overcome by good fertilizer and irrigation management.

Other Constituents

Without going into the details of each and every toxic constituent, it may suffice to add that the problem of toxic elements in water has assumed greater importance. Surface and groundwater are being constantly polluted by industries by directly dumping their wastewater in surface sources of water or groundwater. Problem of groundwater contamination is spreading quite fast due to pollutants leakage from industrial ponds and landfills, septic tanks, mining impoundments and municipal landfills. Besides, chemicals such as herbicides and pesticides used in agriculture are getting into the water resources through runoff after a rain or irrigation event or through seepage into the groundwater. The area underlain with toxic elements is increasing rapidly as more and more analysis results are pouring in. Since the toxic chemicals are a primary cause of cancer, heart disease, strokes, premature aging and birth defects, toxic waters have been in news recently and would continue to dominate the water quality scene in India. The major highlight of this concern however, has been on the direct uptake of toxic water. Rarely, its impact on plants and the entry of toxins in human and animals through the food chain is talked about. Recommended maximum concentrations (mg L^{-1}) of several trace elements in irrigation water are given in Annexure II.

Water Quality for Pressurized Irrigation Systems

Water quality guidelines so far have been prepared for surface irrigation methods mainly because pressurized irrigation methods were not quite common in India until recently. Currently, pressurized irrigation systems cover an area of about 8.5 M ha. With likely increase in the area, guidelines for pressurized irrigation techniques need to be developed. Not much work in India has so far been reported except few observations. Therefore, available guidelines from the literature are collated and reported in this section. Low to medium saline water can be used with sprinkler irrigation. Highly saline water exceeding 8 dS m^{-1} should be used cautiously especially in arid and semi-arid regions during summer season. It may cause leaf burn or deposition of white precipitate on leaves that may affect the photosynthetic activity of the plants. Leaf burn from excess sodium and chloride are known to reduce crop yields (Table 6.11). Bicarbonate rich water can prove to be problematic for drip or sprinkler irrigation systems because of the possibility of calcite or scale build-up causing reduced flow rates through orifices or emitters. Guidelines on susceptibility of clogging in drip irrigation systems due to water quality are reported in Table 6.12 (Nakayama and Bucks, 1991).

Table 6.11 Susceptibility ranges for crops to foliar injury with saline water containing sodium and chlorides

Ion	Na or Cl concentration (mg L^{-1}) causing foliar injury			
	Low	Medium	High	Excessive
Na concentration	<46	46-230	231-460	>460
Cl concentration	<175	175-350	351-700	>700

Table 6.12 Susceptibility of clogging in drip emitters due to water quality

Potential problem	Units	None	Slight to Moderate	Severe
Physical				
Suspended Solids	mg L^{-1}	< 50	50-100	>100
Chemical				
pH		< 7.0	7.0-8.0	> 8.0
Dissolved Solids	mg L^{-1}	< 500	500-2000	>2000
Manganese	mg L^{-1}	< 0.1	0.1-1.5	>1.5
Iron	mg L^{-1}	< 0.1	0.1-1.5	>1.5
Biological				
Bacterial populations	Number per L	<10 000	10 000 – 50 000	>50 000

CLASSIFICATION OF IRRIGATION WATERS

A number of water quality classifications have been developed and reported but most are location and country specific. The USSL diagram has been widely used globally and in India. Several classification criteria are available to categorize water quality specific to the Indian conditions. Most widely used is the one developed by a Committee of Experts and is quite elaborate compared to the USSL diagram. This section limits its discussion to these classification criteria as others are mainly the offshoots of these classifications.

USSL Diagram

The United States Salinity Laboratory (Richards, 1954) published a semi-logarithmic diagram, which rated waters on the basis of its EC and SAR. Four classes each of EC and SAR were devised (Table 6.13, Fig. 6.9). Each salinity class was associated with all the four sodium classes or vice versa, thus giving a total of sixteen classes. It was concluded that low salinity water (C1) can be used for irrigation with most crops on most soils with little likelihood that soil salinity will develop. On the other hand, high salinity water (C4) was characterized as unsuitable for irrigation under ordinary conditions, but might be used occasionally under very special circumstances. The conditions associated with the use of this water are: soil must be permeable, drainage must be adequate, irrigation water must be applied in excess to provide considerable leaching and highly salt tolerant crops should be chosen for cultivation.

Table 6.13 EC and SAR values for different classes of salinity and sodium hazard of irrigation water

Salinity hazard	Class	EC (dS m^{-1})	Sodium hazard	Class	SAR
Low	C1	0.1-0.25	Low	S1	<10
Medium	C2	0.25-0.75	Medium	S2	10-18
High	C3	0.75-2.25	High	S3	18-26
Very high	C4	2.25-5.0	Very high	S4	26-31

The diagram is based upon the dependence of sodium hazard on salinity as a result of which water at the same SAR were given a higher sodium hazard rating with increasing water salinity and vice versa. Thus, permissible limits of SAR decreased from 10 to 8, 6, 4 and 2 as EC increased from 0.1 to 0.25, 0.75, 2.25 and 5.0 dS m^{-1}, respectively to be classed as S1, water having low sodium hazard. All water having SAR of 10 will be either S2 or S3 i.e. medium to high sodium hazard. On the basis of location specific observations, several limitations of the USSL diagram have been pointed out by various workers and users. Some of them may or may not be tenable. Nonetheless, the USSL diagram does not give clear-cut assessment of the quality of irrigation water for the monsoon climatic conditions of India.

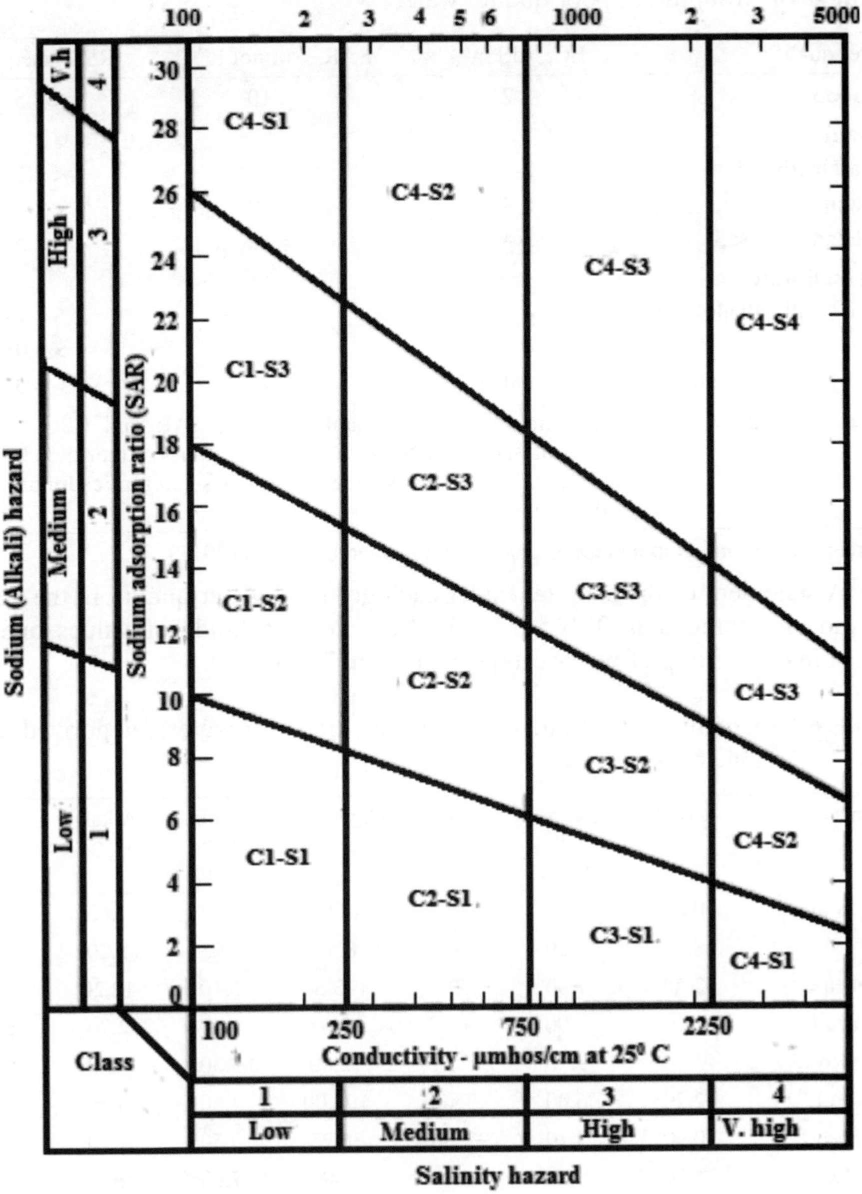

Fig. 6.9 USSL diagram to categorize irrigation water on the basis of EC and SAR

AICRP Classification

Based on EC, SAR, RSC and toxic elements (added later), a Committee of Experts grouped the water into four main classes (Table 6.14). Saline and alkali water have further been sub-grouped into three classes each to grade the severity of the problems. Apparently the classification is considered broad based than any other classification prevalent in India.

Table 6.14 Grouping of poor quality water

Water quality	EC_{iw} (dS m^{-1})	SAR_{iw} (mmol L^{-1})$^{1/2}$	RSC (meq L^{-1})
A. Good	< 2	< 10	< 2.5
B. Saline			
i. Marginally saline	2-4	< 10	< 2.5
ii. Saline	> 4	< 10	< 2.5
iii. High-SAR saline	> 4	> 10	< 2.5
C. Alkali water			
i. Marginally alkali	< 4	< 10	2.5 – 4.0
ii. Alkali	< 4	< 10	> 4.0
iii. High-SAR alkali	variable	> 10	> 4.0
D. Toxic Water	The toxic water has variable salinity, SAR and RSC but has excess of specific ions such as nitrate, boron, fluoride, chloride, sodium, silica or heavy metals such as selenium, cadmium, lead and arsenic etc.		

Source: Minhas and Gupta (1992), Toxic waters added by Gupta (2011)

A statement of the samples under each group of water quality in the state of Haryana is included in Table 6.15. It illustrates the kinds of values one may expect in each group of water category given in Table 6.14.

Table 6.15 Chemical constituents of various kinds on water categorized as per Table 6.14

Parameter	Good	Marginally saline	Saline	High SAR saline	Marginally alkali	Alkali	High SAR alkali
EC (dS m^{-1})	0.61	2.56	6.52	14.04	1.04	1.83	2.53
pH	8.60	8.30	8.3	8.55	8.05	8.20	8.50
Na (meq L^{-1})	2.20	7.00	32.8	103.60	6.10	12.20	21.70
Ca (meq L^{-1})	2.10	3.00	10.3	22.64	1.50	2.70	2.10
Mg (meq L^{-1})	1.70	9.70	20.7	12.60	3.30	3.10	1.70
Cl (meq L^{-1})	2.20	6.00	33.00	102.00	3.60	3.80	6.80
SO_4 (meq L^{-1})	0.90	3.10	25.40	36.75	0.50	1.70	9.80
CO_3+HCO_3 (meq L^{-1})	4.20	13.20	6.00	2.40	7.60	13.00	10.60
SAR (meq L)$^{1/2}$	1.60	2.89	8.33	24.80	3.90	7.20	15.72
RSC (meq L^{-1})	0.40	1.50	Nil	Nil	2.80	7.20	6.80
Mg/Ca ratio	0.81	3.23	2.01	0.56	2.20	1.15	0.81
Cl/SO_4 ratio	2.44	1.94	1.30	2.78	7.20	2.24	0.69
Village	Allipur	Barota	Bindrala	Durjanpur	Alipur	Amunpur	Kabulpur
Block	Nissing	Nissing	Assandh	B. Khera	Nissing	Nissing	Assandh
District	Karnal	Karnal	Karnal	Bhiwani	Karnal	Karnal	Karnal

DETERMINATION OF WATER QUALITY PARAMETERS

Water analysis report of a given water sample is

Ca	=	2.6 meq L^{-1}	Mg	=	1.9 meq L^{-1}
Na	=	21.7 meq L^{-1}	EC	=	2.5 dS m^{-1}
Cl	=	6.3 meq L^{-1}	SO_4	=	9.7 meq L^{-1}
$CO_3 + HCO_3$=		10.6 meq L^{-1}			

Find SAR, RSC, and adj. R_{Na}

Check

Sum of anions and cations is:

$$2.6+1.9+21.7 = 6.3+9.7+10.6$$

$$26.2 \sim 26.6$$

SAR $= Na/\sqrt{(Ca+Mg)/2} = 21.7/\sqrt{(2.6+1.9)/2} = 21.7/\sqrt{2.25} = 32.8/1.5 = 14.5$

RSC $= 10.6 - (2.6+1.9) = 10.6 - 4.5 = 6.1$ meq L^{-1}

Adj. R_{NA} $= Na/\sqrt{(Ca_x+Mg)/2}$

EC is 2.5 dS m^{-1}, HCO_3/Ca ratio 10.6/2.6 = 4.08 (Assuming carbonate as 0.0 meq L^{-1})

Ca_x value from Table 6.7 corresponding to the nearest values of HCO_3/Ca and EC_{iw} is 0.89

Adj. R_{NA} $= 21.7/\sqrt{(0.89+1.9)/2} = 21.7/\sqrt{1.40} = 21.7/1.18 = 18.38$

Apparently on the basis of Adj. R_{NA}, sodicty hazard is more than given by SAR.

Management Strategies

No single strategy can be prescribed when crops are irrigated with saline/alkali waters. Attempts to develop various strategies have yielded a large number of them. All of these can be grouped under the five heads namely

- Crop management
- Soil management
- Irrigation water management
- Chemical management
- Rainwater management

Within these groups, several sub-groups have been identified. These groups and sub-groups are listed in Table 6.16. Actionable activities under each sub-group also form a part of this table. Thus, one can have a holistic view of options available to manage saline/alkali waters.

Table 6.16 Management of saline/alkali water

Group	Technology	Actionable activities
Crop management	Crop selection	Select crops and cropping pattern on the basis of soil salinity/alkalinity likely to be developed with a given quality of irrigation water (Chapter 5)
	Exploitation of varietal differences	Select crop varieties that are more salt tolerant (Chapter 5)
	Growth stage sensitivity	Apply fresh water at salt sensitive growth stages to provide relatively salt free environment
	Agronomic practices	Use 25% higher seed rate; higher plant population in transplanted crops; apply post sowing irrigation, if necessary; utilize stored moisture for crop seeding
Soil management	Soil type specific management	In general light textured soils can be managed better than heavy textured soils
	Land forming/ seeding	Avoid seeding spots where salt accumulate; land forming and seeding practices can help to live with the salts
	Fallowing	In low rainfall region, it may be prudent to fallow the land during monsoon season for soil rejuvenation
Irrigation water management	Shallow depth-high frequency irrigation	Low-depth high frequency irrigation helps to manage saline water, note that to restrict the salt load, total amount of water should remain the same
	Conjunctive use of multi-source water	Conjunctive use of saline and fresh water (either from canal or harvested rain water); it can be achieved either through blending or cyclic use mode
	Pre-sowing irrigation	As far as possible, apply pre-sowing irrigation with fresh water to get higher yields in poor quality water irrigated lands
	Switchover to improved irrigation techniques	Improved irrigation practices such as drip and sprinkler help to adopt shallow depth-high frequency irrigation; since matric stress is minimized, it helps the crops to tolerate higher osmotic stress
	Leaching fractions	Maintain appropriate leaching fractions to restrict salt build-up in the soils
Chemical management	Application of chemical amendments*	Alkali water needs some kind of amendment to neutralize high RSC, high SAR water may also need little dose of amendment to maintain infiltration rates
	Application of additional nutrients	About 25% more nitrogen should be applied to soils irrigated with saline/alkali waters; additional application of phosphorus may be helpful in soils irrigated with chloride rich saline water, zinc sulphate at 10-20 kg ha^{-1} should be applied, particularly to alkali water irrigated lands

Group	Technology	Actionable activities
Rainwater management	*In-situ* rainwater conservation	Rainwater is the best quality water, which should be conserved in saline water irrigated lands to leach the applied salts
	Rainwater harvesting and reuse	In the absence of fresh water, rainwater harvesting and its use in conjunctive use mode with saline water helps to manage saline water

* Described in a separate section

Application of Amendments

The alkali hazard of the irrigation water on soils can be mitigated by neutralization of RSC of the irrigation water by using a chemical amendment. Gypsum is by far the most important chemical amendment. Several other amendments such as acid forming substances (notably pyrites) or industrial wastes such as distillery spent wash (DSW), phosphogypsum (PG) and sulphitation press mud (SPM) have been used as soil application (Anonymous, 2007, Anonymous, 2009, Gupta, 2010, Gupta and Gupta, 2019). The benefits of acid forming amendments are subject to the calcareous nature of the soils. If the RSC of the irrigation water is 2.5 or less, there is no need to add any amendment. However, for the neutralization of every additional 1 meq L^{-1} RSC, agriculture grade gypsum (70% purity) @ 12 kg h^{-1} should be added for each cm of alkali water applied. The quantity of gypsum to be added is, thus determined by the RSC of irrigation water and the quantity of water used in irrigation during the growing season or on yearly basis. Mathematically, amendment requirement to neutralize RSC can be calculated with the help of the following equation.

$$\text{Gypsum to be added (t ha}^{-1}) = 12 \times (\text{RSC}-2.5) \times \text{IW}/1000 \qquad (6.13)$$

Here RSC is the RSC of the irrigation water, IW is the depth of irrigation water applied during the season or year, cm. Division by 1000 is to convert kg ha^{-1} in t ha^{-1}.

Calculation of quantity of gypsum

The water of a tube well has a RSC of 9.5 meq L^{-1}. The water is used to irrigate wheat crop, which needs about 5 irrigations of 6 cm each. Calculate the gypsum requirement for treating the soil to avoid adverse effects of high RSC.

RSC to be neutralized = 9.5-2.5 = 7.0

Irrigation water applied during the season = 5 × 6 = 30 cm

Incorporating in eq. (6.13), we have

$$\text{Gypsum to be added (t ha}^{-1}) = 12 \times 7 \times 30/1000 = 2.52 \text{ t ha}^{-1}$$

Amendments can be applied to the soil either on annual or seasonal basis. Studies on the other hand have revealed that gypsum applied after each irrigation is more effective. However, to apply amendment after each irrigation is cumbersome. Therefore, a gypsum bed technology has been developed, wherein amendment is dissolved in water before it is applied to the field. A schematic and photographic view of the structure is shown in Fig. 6.10.

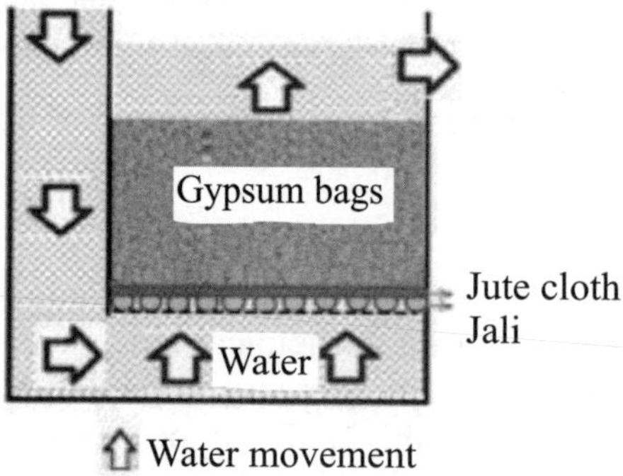

Fig. 6.10 Schematic structure of the gypsum bed (left) and a view of the constructed gypsum bed as it looks in the field (right)

WATR QUALITY AND MANAGEMENT OPTIONS FOR USE

In spite of our expanded knowledge on soil water chemistry, use of poor quality irrigation water has remained an art rather than a science. A major area of concern is that while art is developing, science behind the use has taken a back seat. As such, it has been difficult to generalize and upscale the field results for universal practical applications. The discussion in this chapter has revealed that the degree of adverse effects of irrigation with poor quality water will be governed by the type of crop and its salt tolerance, the characteristics of the soil, soil and water management practices being adopted and climate especially the rainfall and its management. Besides, the effect of salinity and sodicity of irrigation water is quite site specific, making it difficult to set some rigid numbers for irrigation or some other use. Therefore, a stepwise procedure to determine the suitability of poor quality water for irrigation and crop production is proposed in Fig. 6.11. The steps include:

- Collect data on water quality, climate (rainfall), soil properties and irrigation application rates/methods.
- Estimate the leaching fraction and the average root zone salinity/sodicity (Adj. R_{Na}).
- Estimate relative crop yield and compare with the target yield.
- In case of mismatch, one or some combination of the management options is identified to alter the salt and water balance of the root zone.
- Re-estimate relative crop yield and compare with the target yield.
- The process of assessment continues until target yield is achieved.

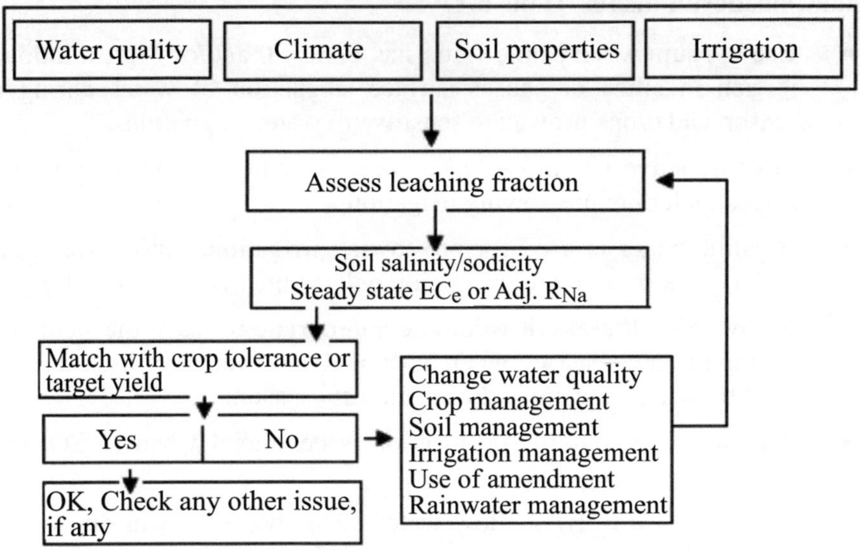

Fig. 6.11 Flow diagram to evaluate salinity and sodicity impacts of irrigation water

GUIDELINES FOR USING POOR QUALITY WATER

Based on the field experience and experimental evidences emerging from all over India, guidelines for using poor quality groundwater have been prepared by a team of experts drawn from AICRP Saline Water, Karnal, CSSRI, Karnal, CCSHAU, Hisar and PAU, Ludhiana (Table 6.17 and Table 6.18).

Table 6.17 Guidelines for using saline waters (RSC < 2.5 meq L^{-1}) for irrigation

Soil texture (% clay)	Crop tolerance	Upper limits of EC_{iw} (dS m^{-1}) in rainfall regions (mm)		
		< 350	350-550	550-750
Fine	S	1.0	1.0	1.5
(<30)	ST	1.5	2.0	3.0
	T	2.0	3.0	4.5
Moderately fine	S	1.5	2.0	2.5
(20 – 30)	ST	2.0	3.0	4.5
	T	4.0	6.0	8.0
Moderately coarse	S	2.0	2.5	3.0
(10 – 20)	ST	4.0	6.0	8.0
	T	6.0	8.0	10.0
Coarse	S	-	3.0	3.0
(< 10)	ST	6.0	7.5	9.0
	T	8.0	10.0	12.5

S, ST and T denote sensitive, semi-tolerant and tolerant crops to salinity of irrigation water

Special Considerations for Table 6.17

- Use gypsum when saline water has SAR>20 and/or Mg/Ca ratio >3 or is rich in silica or causes surface stagnation of water during rainy season and crops grown are sensitive to water stagnation.
- In conjunctive use mode, canal water should be used at early growth stages including pre-sowing irrigation.
- If saline water is used for pre-sowing irrigation, 20% extra seed rate and a quick post-sowing irrigation will ensure better germination.
- In low rainfall areas, it would be appropriate to leave the fields fallow during the rainy season when SAR > 20 and waters of high salinity are used to irrigate the crops during the *rabi* season.
- Additional phosphorus fertilization is beneficial when Cl/SO_4 ratio is >2.0.
- For soils having (i) shallow water table (within 1.5 m during *kharif*) and (ii) hard sub-soil layer, the next lower EC_{iw} /alternate mode of irrigation (canal/saline) is applicable.
- Use of organic materials in saline environment enhances yields.

Comments

i. Textural criteria should be applicable for all soil layers down to at least 1.5 m depth.

ii. In areas where groundwater table reaches within 1.5 m at any time of the year or a hard sub-soil layer is present in the root zone, the limit of the next finer textural class should be used.

iii. If in any textural class, SAR and RSC values are above the upper limits, gypsum should be used as per the requirement.

Table 6.18 Guidelines for using alkali water (RSC > 2.5 meq L^{-1} and EC_{iw} < 4 dS m^{-1}) for irrigation

Soil texture (% clay)	Upper limits of		Remarks
	SAR	RSC	
Fine (>30)	10	2.5 – 3.5	Limits pertain to *kharif* fallow/*rabi* crop rotation when annual rainfall is 350-550 mm
Moderately fine (20-30)	10	3.5 – 5.0	When the water have Na < 75% (Ca + Mg > 25%) or rainfall is < 550 mm, the upper limit of the RSC range become safe
Moderately coarse (10-20)	15	5.0 – 7.5	For double cropping, neutralization of RSC with gypsum is essential
Coarse (< 10)	20	7.5 –10.0	

Source: Minhas and Gupta (1992)

■■■■

Chapter - 7

METHODS OF DETERMINATION: CHEMICAL PROPERTIES

Nature and properties of saline, saline-alkali and alkali soils are quite different than the normal or acid soils. As a result, these soils pose insurmountable problems during cultivation because of poor physical state and/or excessive chemical constituents. To understand the nature and properties of these soils, it is utmost important to have chemical and physical tests performed on representative soil samples. The kind and number of tests required to assess the kinds and degree of problems vary from soil to soil and one kind of assessment to another. The need to test these soils also arises because there is a gradual improvement in soil's physical, chemical and biological properties over the years. The dynamics of these changes depends upon several factors. Tests to measure these changes reveal the current status of the soil and help to decide appropriate management options and cropping patterns. This chapter describes the soil sampling procedure and chemical tests relevant to these soils having bearing on crop productivity. Tests to evaluate physical properties of these soils that impacts crop productivity are included in the next chapter.

CONCEPT OF SOIL TESTING

Soil testing is the rapid chemical analysis of a soil to estimate soil reaction (pH), salinity/alkalinity and available nutrient status of the soil. Soil testing is performed to diagnose the problems and to identify appropriate management interventions/recommendations. A soil testing programme has following 4 phases:

- Collection and preparation of soil samples
- Chemical analysis of soil samples for extraction and determination of relevant physico-chemical properties of the soil
- Interpretation of analytical results
- Recommendation and follow up of results.

DRAWING AND PROCESSING OF SOIL SAMPLES

The axiom '*the analysis can be no better than the sample*' is an apt description of the importance of a soil and water sample used in the analysis to determine either chemical or physical properties of soil or water. For the results to be of any value, it is necessary that the soil samples should be representative of the area. Further, the handling operation in the laboratory should be carefully performed because a very small quantity (1-10 g) of the large soil mass of the field is actually used for the analysis in the laboratory. Unless one is sure of representativeness of samples, one may end up in misleading interpretation and drawing wrong recommendations.

Apparatus and Materials: *Khurpi,* Spade, Augers (tube, post-hole or screw-type auger), Plastic bowl or trays, Scale, Rack, Wooden roller, Mortar and pestle, Sieve, Polythene/paper/cloth bags, Labels, Card board cartons, Aluminium boxes/cans

It may be useful to divide the area in different homogenous units on the basis of differences in soil type, colour, crop growth or slope. In a uniform field, demarcate the sampling points in a zigzag fashion or distribute the locations randomly in such a way that the whole field is covered in about 8 samples per acre or 20 sample per ha (Gupta et al., 2012). For diagnostic purpose, one sample of top 0-15 cm will suffice although from management point of view, two samples of 0-15 and 15-30 cm depth should be drawn. For saline soils another sample from 30-60 cm may also be taken.

Appropriate Time of Sampling

The time of sampling mainly depends upon the purpose for which samples are being drawn. Soil samples can be drawn at any time provided the soil conditions permit. For diagnosis of salt affected soils in India, the time period after February to June is most suitable. If some cropping is being done, then the ideal time is just after the harvest of the *Rabi* crops. The samples may not be drawn during or immediately after the monsoon season. It is important to sample at similar times year after year so that it is easier to compare results at regular time intervals.

Drawing Primary Samples

While taking samples, remove the surface litter with *Khurpi* or spade. In salt affected soils, a salt crust or salt efflorescence may be encountered at the soil surface. A thin slice of this layer may be removed before sampling. With the help of an appropriate sampling tool (Auger), collect a sample in a plastic tray or on a clean bag. For a moist soil, the tube auger or post hole auger or spade may prove satisfactory. For harder soil, a screw auger may be more convenient. If augers are not available, dig a V-shaped hole up to the depth of sampling. Then take a fine thick slice from the smoothened side using a spade. Trim the sides leaving a fine strip in the middle, being the sample. The samples so drawn are known as 'primary' samples. Each primary sample should be approximately of the same weight.

Following precautions should be taken during collection of soil samples.

- Do not sample unusual areas, such as marshy areas, old paths, old channels, old bunds, areas near trees, and other unrepresentative sites.

- Avoid sampling from sites where manure has been dumped or fertilizer bags have been stacked. Also avoid spots where farm residue has been burnt.
- Take separate sample from the areas of different appearances. In row crop take sample in between rows.
- Keep the sample in a clean bag.
- If the soil is to be used to estimate micro-nutrients such as Zn, Cu, Fe, Mn, use of metallic tools should be avoided. Samples must be collected by steel or rust free *khurpi*/auger and kept in clean polythene bag. Brass sieve should be avoided. Nylon net or aluminium sieve should be used.
- Use clean bags for sample collection. Do not use bags which had earlier contained fertilizer, manure or plant protection chemicals etc.
- Avoid contamination of the samples at any stage of handling.

Compositing Samples

After collecting the required number of samples, thoroughly mix all the samples for the area and for the same depth. While mixing, break the clods, remove large rocks, pieces of sod and roots, earthworms, etc. Since about 0.5 kg sample is required for analysis, composite sample is prepared and reduced in size by quartering method. In the quartering method, the sample is approximately divided into 4 equal parts. While 2 opposite portions of the samples are discarded, remaining 2 portions are retained. If the weight of the sample is still more, thoroughly mix the retained parts and again divide into 4 equal parts and discarding 2 opposite parts again. This procedure is continued until 0.5 kg sample remains. It is known as composite sample which is considered to be the true representative of the area.

The soil samples should be packed in polythene bags of 6”×9’’, which may be sealed by twisting or tying the neck or by means of rubber bands or adhesive tape. Soil samples are labeled/numbered with water proof ink or paint. These bag numbers are entered in the sampler’s record book, along with other relevant information supplied by the stakeholders. If labels are used, they should have either printed numbers on them or water proof ink should be used to write information on the labels. For soil profile samples, depth of sampling may be mentioned on the bag.

On arrival of the soil samples at the laboratory, the content of a box should be checked against the packing note prepared by the sampling team. Any discrepancies should be immediately communicated to the sampler. It is essential to keep a record of the date of arrival and the source of all samples. A table can be prepared date wise for each month. The information on the soil sample such as sample number, name and address of the farmers, details of the field and site, local name of field, *Khasra* number etc. must be legibly and correctly noted in the register.

Sampling Slick Spots

The slick spots in salt affected soils can best be identified during the cropping season. It is better to sample when spots are visible. Samples representing slick spots should not be mixed with samples from good spots. If salt crust is sampled, it should not be mixed with the soil samples and separately kept for analysis.

Preparation of Soil Sample

- Discard large stones and pieces of gravel. Remove roots, leaves and other un-decomposed organic residues from the samples.
- Large lumps of moist soils should be broken by hand. Spread out the soil in a thin layer on clean cloth, plastic or brown paper sheet for drying.
- The soil samples are air dried in shade at room temperature. The soil chemical reactions in dried soils are at nearly equilibrium.
- After air drying, the samples are crushed gently using mortar and pestle and sieved through a 2 mm sieve. The coarse portion on the sieve is returned to the mortar for further grinding. Pebbles, concretions, and stones should not be broken during grinding. Repeat sieving and grinding till all aggregate particles are fine enough to pass the 2-mm sieve. However, for analysis of carbon, nitrogen, and total elements (arsenic, cadmium, copper, chromium, iron, lead, manganese, molybdenum, nickel, phosphorus, selenium, zinc), samples should be grounded to a fine powder that passes through a 60-mesh screen.
- About 250 g of sieved sample is kept in properly labeled sample bag for testing.
- The registered and labeled samples are finally placed in a cardboard carton. Label the carton properly with the details of soil sample and store in a separate room. The room should not be exposed to heat or cold or dampness.

DRAWING OF WATER SAMPLES

Proper diagnosis of the nature and the amount of salts present in water is essential to assess the possible hazards which it may pose to the soil or the crop. Collection of representative sample is, therefore, basic for correct diagnosis and for making sound recommendations for the management of poor quality waters. Following procedures is adopted to collect a water sample.

In case of gravity canal, take the sample of the water from the channel after irrigation has been in progress for about 15 min to half an hour. Groundwater sample should be drawn only after the tube well has run for about 30 min. As far as possible, plastic bottles should be used to collect water samples. Glass bottles can be used, if plastic bottles are not available. Avoid using metal containers to collect water samples. Bottles should be thoroughly cleaned and rinsed with the source water 2-3 times before taking the samples. For most routine salinity appraisal tests, about 100 mL of water will be sufficient. To be on the safer side and to allow specialized tests, if needed, a minimum of 500 mL sample should be taken.

Water containers should be properly screwed and if possible wax sealed. Containers should be properly labeled for identification. Each sample should be accompanied by appropriate description giving name of the farmer, village, source of water, type of tube well for groundwater, and crops being grown etc. It will help the laboratory to fine-tune recommendations

In recent times, water quality is getting more and more attention. Agricultural scientists are often called upon to test trace elements, toxic constituents and waste water quality. In these cases one has to take special care in sampling, preservation and storage of samples. Some guidelines are given in Table 7.1.

Table 7.1 Guidelines for sampling of water

General physico-chemical parameters, anions and cations, all forms of nitrogen and phosphorus	Plastic bottles, with or without air	No additive is needed	Temperature and dissolved oxygen should be measured at site
Chemical Oxygen Demand (COD)	Plastic bottles, no air	Sulphuric acid	Additive is not required if samples are analyzed within 48 h
Biological Oxygen Demand (BOD)	Plastic bottles, no air	No additive	
Trace elements	Plastic bottles, with or without air	Nitric acid	Except mercury (Hg)
Trace organics and pesticides	Dark glass bottle, no air	No additive	
Microbiological parameters (total and faecal coliforms, helminths, viruses, etc.)	Sterile plastic bottle, with air	No additive	Sodium thiosulphate in presence of residual chlorine should be added to disinfect the effluent

PREPARATION OF SOIL SATURATION PASTE

A saturation extract is routinely used to determine chemical properties of salt affected soils. The cations analyzed in saturation extracts are Ca^{2+}, Mg^{2+}, K^+, and Na^+, while the anions are SO_4^{2+} (sulphate), CO_3^{2+} (carbonate), HCO_3^- (bicarbonate) and Cl^- (chloride). Boron (B) in saturation extracts is often measured where its toxicity is suspected.

Apparatus: Porcelain dishes, Spatulas or mixing spoons, Vacuum filtration system

Procedure

- Weigh 200-300 g air-dry soil (< 2 mm) into a porcelain dish.
- Slowly add distilled water, and mix with a spatula until the soil paste does not have free-standing water on the surface of the paste. It slides freely and cleanly comes off a spatula (does not apply to clay soils, > 40% clay). It flows slightly when the container is tipped to a 45° from horizontal. Soil surface glistens as it reflects light. It consolidates easily by tapping after a cut is formed in the paste with the flat side of a spatula (does not apply to sandy soils, >70% sand).
- Allow the paste to stand for 1 hour, and then re-check the criteria for saturation by either adding more distilled water or soil, as per requirement.
- Leave the paste for 6 to 16 hours or better overnight. For salinity appraisal, the saturated paste can be extracted after 4 hours. For alkali soil samples, it should stand 16 or more hours. For the soluble B, it should stand 24 hours or more for complete equilibration. Re-check the criteria for saturation by adding more distilled water or soil, as needed. Filter with a vacuum filtration system using a Buchner funnel fitted with Whatman No. 42 filter paper.
- Collect the filtrate in a bottle and use it for analysis.
- If the initial filtrate is turbid, filter the extract.

CHEMICAL PROPERTIES

Total Dissolved Solids (TDS)

TDS comprises of inorganic salts (principally calcium, magnesium, potassium, sodium, bicarbonates, chlorides and sulphates) and small amounts of organic matter that might get dissolved in water. Although the determination of TDS in agricultural sector is becoming obsolete yet in many other fields the term is still in vogue. The two principal methods of measuring total dissolved solids are gravimetric analysis and electrical conductivity. When EC of the saturation extract of the soil or water is known, following equations can be used to calculate the TDS.

$$\text{TDS (mg L}^{-1}) = 640 \times \text{EC (dS m}^{-1}) \text{ (EC} < 5 \text{ dS m}^{-1}) \quad (7.1)$$

$$\text{TDS (mg L}^{-1}) = 800 \times \text{EC (dS m}^{-1}) \text{ (EC} > 5 \text{ dS m}^{-1}) \quad (7.2)$$

The gravimetric analysis is done as per the following procedure.

Apparatus/Reagents: Platinum dish, Whatman filter paper

Procedure

Filter the sample to obtain a perfectly clear liquid. Choose sample volume to yield between 10 and 200 mg dried residue. Transfer this aliquot in a weighed platinum dish and evaporate to dryness. Dry to constant weight at 105°C or in an oven at 180°C. Cool in a desiccator and weigh.

Calculations

$$\text{Dissolved solids (ppm)} = (W_{dr} - W_d) \times 10^6 \times mL^{-1} \text{ of sample used} \quad (7.3)$$

Here W_{dr} is the weight of the dish with residue, g and W_d is the weight of the dish without residue, g.

Electrical Conductivity (EC)

Apparatus/Reagents: Conductivity bridge (Fig. 7.1) with conductivity cell, Standard potassium chloride solution (0.02 N KCl). It is prepared by dissolving 1.4912 g of KCl in water and making the volume to 1 L. It has an electrical conductivity of 2.768 dS m^{-1} at 25°C.

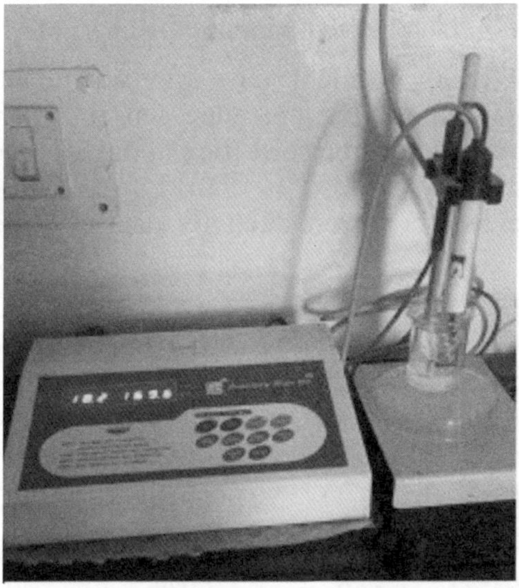

Fig. 7.1 A view of the EC meter with temperature sensor

Procedure

1. Pour the standard solution (EC =2.768 dS m^{-1}) into a 100 mL beaker
2. Immerse the conductivity cell (previously rinsed with the standard solution) into the beaker. Note down the conductivity as EC_m.
3. Determine the cell constant using the following relation.

$$\text{Cell constant (k)} = EC_a/EC_m \quad (7.4)$$

Here , EC_a is the actual EC (2.768 dS m^{-1}) and EC_m is the measured EC in dS m^{-1}.

4. Take out the conductivity cell and rinse the same with distilled water. Use the test solution (soil saturation extract or water) to measure the

conductance as before. The actual EC of the test solution is then given as:

$$EC = \text{Measured EC} \times k \quad (7.5)$$

Temperature correction

Since EC is commonly reported at 25°C, temperature of the test solution is also taken. For this purpose, most laboratory meters are provided with temperature correction. Tables are also available to convert EC measured at a given temperature to the EC at 25ºC (Table 7.2). As a thumb rule, the electrical conductivity increases approximately by 2% per degree C rise in temperature. For example to convert values measured at 20°C to the values at 25°C, multiply by 1.112. However, use of table is recommended such that

$$\text{EC at } 25°C = \text{EC at } T°C \times f_t \quad (7.6)$$

Here, f_t = correction factor at temperature t from Table 7.2.

Note: The new EC meter now available have provision to automatically take care of the temperature correction. A temperature sensor attached to the electrode senses the temperature for automatic correction in the EC value.

Table 7.2 Correction factors for conductivity data at standard temperature of 25°C

°C	f_t	°C	f_t	°C	f_t
08.0	1.488	22.6	1.051	29.2	0.921
09.0	1.448	22.8	1.047	29.4	0.918
10.0	1.411	23.0	1.043	29.6	0.914
11.0	1.375	23.2	1.038	29.8	0.911
12.0	1.341	23.4	1.034	30.0	0.907
13.0	1.309	23.6	1.029	30.2	0.904
14.0	1.277	23.8	1.025	30.4	0.901
15.0	1.247	24.0	1.020	30.6	0.897
16.0	1.218	24.2	1.016	30.8	0.894
17.0	1.189	24.4	1.012	31.0	0.890
18.0	1.163	24.6	1.008	31.2	0.887
18.2	1.157	24.8	1.004	31.4	0.884
18.4	1.152	25.0	1.000	31.6	0.880
18.6	1.147	25.2	0.996	31.8	0.877
18.8	1.142	25.4	0.992	32.0	0.873
19.0	1.136	25.6	0.988	32.2	0.870
19.2	1.131	25.8	0.983	32.4	0.867
19.4	1.127	26.0	0.979	32.6	0.864

°C	f_t	°C	f_t	°C	f_t
19.6	1.122	26.2	0.975	32.8	0.861
19.8	1.117	26.4	0.971	33.0	0.858
20.0	1.112	26.6	0.967	34.0	0.843
20.2	1.107	26.8	0.964	35.0	0.829
20.4	1.102	27.0	0.960	36.0	0.815
20.6	1.097	27.2	0.956	37.0	0.801
20.8	1.092	27.4	0.953	380	0.788
21.0	1.087	27.6	0.950	39.0	0.755
21.2	1.082	27.8	0.947	40.0	0.763
21.4	1.078	28.0	0.943	41.0	0.750
21.6	1.073	28.2	0.940	42.0	0.739
21.8	1.068	28.4	0.936	43.0	0.727
22.0	1.064	28.6	0.932	44.0	0.716
22.2	1.060	28.8	0.929	45.0	0.705
22.4	1.055	29.0	0.925	46.0	0.694

Source: Richards (1954)

EC of soil-water suspensions

Use of saturation extract is recommended to measure the EC of the soil samples. It is customary to denote the soil EC as EC_e, EC of the soil saturation extract. However, in land reclamation works, where numbers of samples are large and analysis is repeated over the years, EC is normally determined in 1:2 to 1:5 soil water suspensions depending upon the soil texture, the higher values being for heavy textured soils. The EC of the soil water paste is denoted as EC_p where subscript p denotes the soil water suspension ratio. For 1:2 soil water suspension EC is denoted as EC_2. Sometimes, it is also reported as $EC_{1:2}$.

To prepare 1:2 soil water suspensions, weigh 25 g of the soil sample and add 50 mL of distilled water in a 100 mL beaker. Stir for 30 min with a glass rod intermittently and wait for a while for the soil particles to settle down. EC of the supernatant solution is measured using EC meter. The EC can be reported as such as EC_2 or can be converted to EC_e using the following relation.

$$EC_e = (\theta_2/\theta_{sp}) \times EC_2 \tag{7.7}$$

Here θ_2 and θ_{sp} are the soil moistures contents in soil-water suspension and at saturation respectively for a given soil.

More often, a standard curve is prepared by analyzing a good number of samples from the project area at the beginning of a land reclamation programme. For these samples, both EC_e and EC_2 are measured. A best-fit line is drawn for the data set, which then can be used to report EC_e values for the soil samples

(Fig. 7.2). The data in Fig. 7.2 reveals that EC_e (Y) is nearly 4.18 times the EC_2 (X).

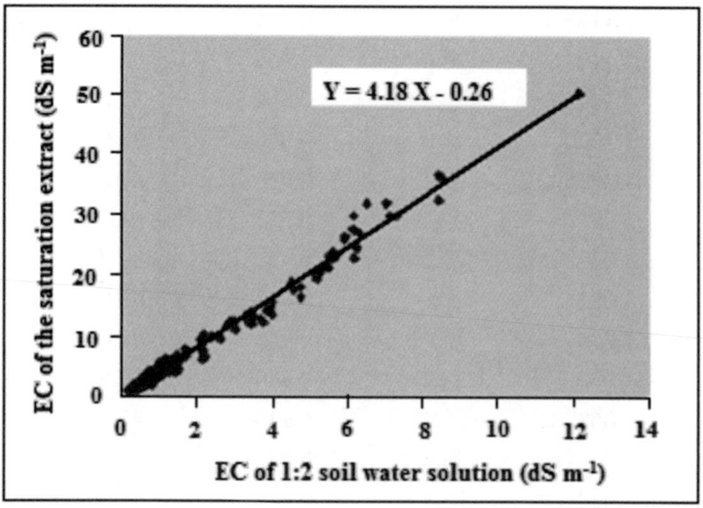

Fig. 7.2 Relationship between EC of the saturation extract and 1:2 soil water solution

Note: Follow all the instructions in the manufacturer's instruction manual for use of the meter and storage of the conductivity cell after use.

pH

The pH of the soil water suspension or extract is an indicator of the nature of the soil whether it is acidic or alkali. A pH less than 7.0 indicates acidic nature while pH above 7.0 indicates the alkali nature.

Apparatus/Reagents: pH meter with glass electrode (Fig. 7.3), Buffer tablets (4 & 7 or 7 & 9.2)

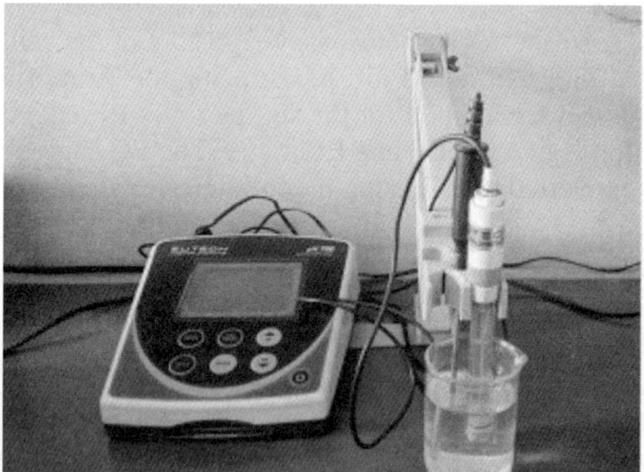

Fig. 7.3 A view of a pH meter

Calibration of pH meter and measurements

1. Wash the electrodes with distilled water using a wash bottle. Dry the electrodes with tissue paper.
2. Use buffer solution of pH 4.0 and calibrate the pH meter to 4.0, the pH of the buffer solution.
3. Repeat the determination with the other buffer solution (say pH 7.0). Repeat these determinations till the pH meter is correctly set to both the buffer solutions. For alkali range, it is advisable to calibrate the pH meter with buffers of pH 7.0 and 9.2.
4. The pH of the water or the soil-water suspension is then read straight from the pH meter.
5. The buffer checks should be repeated in between to ensure that the pH meter remains appropriately calibrated.

Soil pH can be measured using saturation extract or saturation paste or some soil and water ratios say 1:2 or 1:5.

Saturation extract (pH_e)

Use the soil saturation extract to measure pH. It is denoted as pH_e.

pH in saturated soil paste

Add small amount of distilled water to air dried soil. Stir the mixture with a spatula. At saturation, the soil paste glistens and flows slightly when the container is tipped it slides freely and cleanly comes off the spatula. After mixing, allow the sample to stand for an hour. If the paste has stiffened markedly or lost its glistening, add more water or if free water has collected on the surface of the paste, add an additional weighed quantity of dry soil and mix it again. Then insert the electrode carefully in the paste and measure the pH. It is denoted as pH_s.

Measurement of pH in soil to water ratio of 1:2 (pH_2)

Take 20 g soil in 100 mL beaker and add 40 mL of distilled water to it. The suspension is stirred at a regular interval for 30 minutes. Determine the pH by immersing electrodes in suspension. For soils containing high salts, the pH should be determined by using 0.01*M* calcium chloride solution. It is prepared by dissolving 0.110 g of $CaCl_2$ in water and diluted to 1 L. Usually pH_s values are lower than pH_2 by 0.2 to 0.5 unit in normal and saline soils and by 0.5 to 1.0 units in alkali and saline alkali soils due to hydrolysis upon dilution.

Note: Follow all the instructions in the manufacturer's instruction manual for use of the meter and storage of the electrodes.

Determination of Sodium (Richards, 1954)

The Na is determined directly with the help of a flame photometer at emission wavelength of 589 nm (Table 7.3) using appropriate filters and standard curves.

Standard curves are prepared by taking known concentration of Na and plotting the flame photometer readings on Y-axis and known sodium concentrations on X-axis.

Table 7.3 Flame colour and emission wavelengths for various elements

Element	Emission wavelength (nm)	Flame colour
Sodium	589	Yellow
Potassium	766	Violet
Barium	554	Lime green
Calcium*	662	Orange
Lithium	670	Red

*Calcium is measured using the calcium hydroxide band emission

Apparatus and Reagents: Flame photometer, Sodium filter, Volumetric flasks, Beakers, NaCl (AR grade)

Procedure to prepare standard curve

1. Dry NaCl at 105°C for one hour and then cool it in a desiccator. Take 2.5413 g of NaCl, dissolve in water and make the volume to 1 L. It gives a solution of 1000 mg Na mL^{-1}. From this solution take 100 mL and dilute to 1 L to obtain 100 mg Na L^{-1} as stock solution.
2. For preparing working standards, take 0, 5, 10, 15 and 20 mL of stock solution in 100 mL volumetric flask and make up the volume. It will give 0, 5, 10, 15 and 20 mg Na L^{-1}.
3. Place the sodium filter in the flame photometer. Set the milli-ammeter to zero on the zero standard and to full scale deflection with the 10 mg Na standard. Read all of the standards and prepare a standard curve from these measurements taking absorbance on Y-axis and respective concentrations of Na on X-axis.

Test samples

1. Soil saturation extract/water samples are fed on the flame photometer and absorbance is recorded for milli-ammeter readings.
2. Concentration of Na is read on the X-axis against the absorbance on the Y-axis. The values are directly given as mg L^{-1} of Na.

Note: If a water sample is diluted before estimation, the quantity of sodium as observed on standard curve is multiplied by the dilution factor. Where, dilution factor is given as:

$$\text{Dilution factor} = (\text{mL sample} + \text{mL distilled water})/\text{mL sample} \qquad (7.8)$$

Note: Flame photometers currently manufactured obviates the need to prepare calibration curve. Calibration of the flame photometer is done with the help of working standards of 10, 25, 50 and 100 ppm Na. The software prepares the curve and uses the same to display the Na content of the unknown samples. Concentration of Na in ppm is directly read on the LED screen of the flame photometer.

Determination of Potassium (Richards, 1954)

Apparatus and Reagents: Flame photometer, Potassium filter, Volumetric flasks, Beakers, KCl (AR grade), 1 N ammonium acetate solution.

Preparation of standard curve

Dissolve 1.9066 g of dried potassium chloride (AR) in distilled water and dilute to one liter. This is 1000 mg K L^{-1} solution. Take 100 mL of this solution and dilute to one liter to make100 mgK L^{-1} stock solution.

Take 0, 1, 2 3, 4, 5, 6, 7, 8, 9 and 10 mL of 100 mg K L^{-1} solution in different 25 mL volumetric flasks. Make up the volume with 1 N ammonium acetate solution. Adjust the flame photometer reading at zero with blank solution and 100 for 40 mg K L^{-1} solution. Take the flame photometer reading for every dilution using potassium filter at wavelength of 766 nm (Table 7.3). Plot the curve on graph paper by taking K concentration on the X-axis and flame photometer reading on Y-axis. This curve will appear almost similar to that of sodium.

Test samples

1. Soil saturation extract/water samples are fed on the flame photometer and absorbance is recorded for milli-ammeter readings.
2. Concentration of K is read on the X-axis against the absorbance on the Y-axis. The values are directly given as mg L^{-1} of K.

Note: Flame photometers currently manufactured obviates the need to prepare calibration curve. Calibration of the flame photometer is done with the help of working standards of 10, 25, 50 and 100 ppm K. Thereafter, soil saturation extract/water samples are fed on the flame photometer. Concentration of K in ppm is directly read on the LED screen of the flame photometer.

Calcium plus Magnesium (Richards, 1954)

Apparatus/Reagents: Micro- burette, Pipette, Magnetic needles, Stirring plate, China clay dish.

- **Ammonium chloride-ammonium hydroxide buffer solution**: Diss-olve 67.5 g of NH_4Cl in 570 mL of concentrated NH_4OH and dilute to 1L with distilled water. The pH of the buffer is adjusted to 10.0.
- **Sodium hydroxide, 4*N***: Dissolve 160 g of NaOH in 1 L of water. Store in plastic bottle.

- **Standard $CaCl_2$ solution, 0.01*N***: Dissolve 0.500 g of $CaCO_3$, dried at 150°C in 10 mL of approximately 3N HCl and dilute to a volume of exactly 1 L.
- **Eriochrome Black T (EBT) indicator**: Dissolve 0.50 g of Eriochrome Black T and 4.5 g of hydroxylamine hydrochloride in 100 mL of 95% ethanol.
- **Calcon indicator**: Dissolve 0.5 g of calcon in 100 mL of 95% ethanol.
- **EDTA (Di-sodium salt), 0.01*N***: Dissolve 2.0 g of ethylene diamine-tetra-acetic acid-disodium salt and 0.05 g $MgCl_2.6H_2O$ in distilled water, and dilute the solution to 1 L. This is standardized by titrating against 0.01 *N* $CaCl_2$ solution prepared by weighing 0.500 g AR grade $CaCO_3$ (oven dried) and dissolving in minimum excess of dilute HCl (AR) followed by making up the volume to 1 L with distilled water.
- **Polyvinyl alcohol**: 1% solution.
- **Triethanolamine**

Ca in saturated extract

1. Pipette 5 to 25 mL aliquot of the extract (usually 5 mL) containing not more than 0.1 meq of Ca into a china clay dish and dilute to a volume of approximately 25 mL.
2. Add 0.25 mL (5 drops) of reagent 4N NaOH and 4-5 drops of Calcon indicator. Titrate with EDTA, using 10 mL micro-burette. The color changes from pink to blue. When close to the end point, EDTA should be added at the rate of about one drop every 5 to 10 seconds, as the color change is not instantaneous.
3. Carryout a blank in the same manner.

Determination of Ca + Mg

1. Pipette 5 to 25 mL aliquot of the extract (usually 5 mL) containing not more than 0.1 meq of Ca plus Mg in a china clay dish and dilute to approximately 25 mL with distilled water. Add 0.5 mL (10 drops) of ammonium chloride-ammonium hydroxide buffer solution and 3 or 4 drops of eriochrome black indicator.
2. Titrate with EDTA, using 10 mL micro burette. The color changes from wine red to blue or green. No tinge of the wine red color should remain at the end point.
3. Carryout a blank in the same manner.

Or

1. Transfer 5 mL of the saturated paste extract/water sample in 100 mL conical flask and dilute it with distilled water and add 5-10 drops of NH_4Cl - NH_4OH buffer solution.

2. Add a pinch of EBT indicator and titrate it with 0.01 *N* EDTA till the color changes from wine red to blue green.
3. Record the reading

$$\text{Ca or (Ca + Mg)} = (\text{T} \times \text{normality of EDTA solution} \times 1000) / \text{Aliquot (mL) taken} \quad (7.9)$$

Where Ca or Ca+Mg are in meq L^{-1} and T is the volume in mL of standard EDTA used in titration minus the blank.

Mg is calculated from the difference between the (Ca+Mg) and the Ca.

Ca or (Ca + Mg) per 100 gm soil is then given as:

$$(100\times \text{Extract volume (mL)} \times \text{Ca or (Ca + Mg) meq L}^{-1}) / (\text{soil weight x } 1000) \quad (7.10)$$

Carbonates and Bicarbonates (Richards, 1954)

Apparatus/Reagents: Phenolphthalein indicator (1% in 60% ethyl alcohol), Methyl orange indicator (0. 01% in water), 0.01*N* Sulphuric acid, H_2SO_4

Procedure

1. Take 5 mL of the saturated paste extract or water sample in a glass flask. Add 1 or 2 drops of phenolphthalein indicator.
2. If the solution turns pink, titrate against H_2SO_4 until the colour just disappears. Note the volume of the acid consumed and designate it as N.
3. Add 2 drops of methyl orange indicator to the colourless content after the first titration. Again titrate it with H_2SO_4 to the first orange colour and designate the reading as M.

Calculations

$$\text{Carbonates} = (2\text{ N} \times \text{Normality of } H_2SO_4 \times 1000)/ \text{ mL of aliquot taken} \quad (7.11)$$

$$(2\text{ N} \times 0.01 \times 1000) / 5 = 4\text{ N meq L}^{-1}$$

Where, N is amount of reagent used till the end point where pink colour disappears.

$$\text{Bicarbonate} = (\text{M- 2N}) \times (\text{Normality of } H_2SO_4 \times 1000) / \text{ mL of sample taken} \quad (7.12)$$

$$= 2\ (\text{M} - 2\text{N})\ \text{meq L}^{-1}$$

Where, M is amount of reagent used until the end point.

Note: The titrated contents after this analysis can be saved for the analysis of chloride.

Chloride

Mohr's titration method, researched and published by Karl Friedrich Mohr in 1856, is commonly used to estimate chloride. It depends upon the formation of a sparingly soluble brick-red silver chromate ($AgCrO_4$) precipitate at the end point when the sample is titrated against standard silver nitrate ($AgNO_3$) solution in the presence of potassium chromate (K_2CrO_4) as indicator. Initially the Cl ions are precipitated as AgCl and dark brick-red precipitate of Ag_2CrO_4 starts just after the precipitation of AgCl is over.

Apparatus/Reagents: Beakers/porcelain dish, Burette

- **Potassium chromate**, 5% solution: Dissolve 5 g of K_2CrO_4 in about 75 mL distilled water and add drop by drop saturated solution of $AgNO_3$ until a slight permanent red precipitate is formed. Filter and dilute to 100 mL.
- **Standard silver nitrate solution (0.05N)**: Dissolve 8.494 g of silver nitrate ($AgNO_3$) in distilled water and make the volume to one liter. Standardize it against standard NaCl solution and keep in a amber coloured bottle away from light.

Procedure

1. Take 5 mL of the saturated paste extract or water sample in a 100 mL beaker or a porcelain dish and dilute to 25 mL with distilled water. Sample preserved from the carbonate, bicarbonate determination can also be used.
2. Add 5-6 drops of K_2CrO_4 indicator (making it dark yellow). Titrate against the standard $AgNO_3$ solution with continuous stirring till the first brick-red tinge appears.
3. It is usually recommended to determine a blank correction by titrating a like volume of Cl free distilled water.

$$\text{Cl (meq L}^{-1}) = [(\text{mL of AgNO}_3 - \text{Blank}) \times 0.05*1000] / \text{mL of aliquot taken} \quad (7.13)$$

Determination of Chloride with Chloridometer

Principles of Operation

The silver and chloride ions combine in quantitative reactions that result in an insoluble precipitate of silver chloride (AgCl). This reaction is carried out at a constant rate by passing a fixed direct current between a pair of silver electrodes immersed in an acid solution. The anode, which is consumed in this reaction, is a continuous spool of silver wire. As soon as the immersed portion of the wire is used up in the reaction, fresh wire is drawn out from the spool. At the equivalence point of the reaction, the current flowing between a pair of separate indicator electrodes increases and is detected by the sense circuitry. At a pre-set indicator current, the instrument automatically stops the increment in counter and the generation of silver ions. Since the generator current is constant, the total titration time is directly proportional to the number of chloride ions that are introduced into the sample vial. The instrument displays this relative time in units of meq L^{-1} of chloride.

Apparatus/Reagents: Chloridometer

- **Chloridometer Acid Reagent**: Prepare this reagent which contains 0.6 N HNO_3 and 60% CH_3COOH by taking 350 mL of distilled water and 36 mL of concentrated nitric acid and 600 mL of glacial acetic acid and make the volume to one liter.
- **Gelatin Indicator**: Store after grinding the following in a mixer:
- Gelatin – 60 g
- Thymol – 1 g
- Thymol blue – 1 g

Take from this mixture 0.62 g and dissolve this in 100 mL of boiling distilled water.

- **Standard solution of Chloride**: Dissolve 0.585 g of NaCl in a liter of water to obtain 10 meq L^{-1} of chloride solution.

Procedure

1. To standardize the instrument, take 1 mL of 10 meq L^{-1} chloride solution in a vial. Add 4 mL of distilled water and 1 mL of Chloridometer Acid Reagent. Six drops of gelatin indicator are also added to the same vial.
2. Turn on the instrument, clean the electrodes and put the vial in the sliding housing. Slide the vial so that the electrodes and its stirrer dip in the vial having standard solution. Titration will begin automatically and the display will show the progress of titration. At the end point, the displaying digital counter will stop. Note the reading for 10 meq L^{-1} of chloride solution.

   Reading for 10 meq L^{-1} of chloride = Y (say)

   Factor to convert reading into meq L^{-1} of chloride = Y/10 = Z

Now take 1 mL of the sample in another vial. Add 4 mL of distilled water and 1 mL of Chloridometer acid reagent. Also add 6 drops of gelatin indicator. At the end point, note the reading. At this point displaying digital counter will stop.

$$\text{Chloride (meq L}^{-1}\text{)} = \text{Reading} / Z \qquad (7.14)$$

Note: Chloridometer method is used only if chloride content in the samples is expected to be in trace amounts.

Sulphate

Sulphate is determined by number of methods such as gravimetric, colorimetric, turbidimetric or titrimetric methods. The turbidimetric method measures the sulphate content by the extent of turbidity created by precipitated colloidal barium sulphate suspension. Degree of turbidity is measured by turbidity meter or estimated by spectrophotometer at 440 nm.

Apparatus and Reagents: Turbidity meter, Spectrophotometer, Beakers, Volumetric flasks

- **Sodium acetate-acetic acid buffer**: Dissolve 100 g of pure sodium acetate in 200 mL of distilled water. Add 31 mL of glacial acetic acid and make the volume to 1 L. Adjust pH to 4.8.
- **Gum acacia**: Dissolve 2.5 g of gum acacia in one liter of distilled water. Keep it overnight and filter.
- **Barium chloride**: Pure $BaCl_2$ crystals ground to pass through 0.5 mm sieve but retained on a 0.25 mm sieve.
- **Potassium sulphate solution**: To make a stock solution of 10 meq S L^{-1}, weigh 1.74 g of pure K_2SO_4 and dissolve in one liter water.

Procedure

1. To prepare the standard curve, pipette 0, 2.5, 5, 7.5, 10 and 12.5 mL from stock solution containing 10 meq S L^{-1} into 25 mL volumetric flasks. Bring the contents to volume so that the mixture in flask contains 0, 1, 2, 3, 4 and 5 meq S L^{-1}. Then develop the turbidity as illustrated in point 3 and 4. Measure the turbidity. Draw a curve with sulphur concentration on the X-axis and absorbance on the Y-axis. It is used to measure the unknown sulphur content in the samples.
2. Take 5 mL of the water sample (having < 1 meq S L^{-1}) in 25 mL volumetric flask. If the EC of water is >1 dS m^{-1}, dilute it with distilled water to bring EC below 1 dS m^{-1}.
3. Add 10 mL of sodium acetate-acetic acid buffer to maintain the pH around 4.8. Add 1 mL of gum acacia and 1 g of $BaCl_2$ crystals and shake well. Make the volume to 25 mL with distilled water.
4. Invert the flask several times and measure the turbidity with a spectrophotometer at 440 nm using blue filter.

Calculation

Calculate the S content of samples using the standard curve. Multiply it by the dilution factor 5 (5 mL made to 25 mL) express the results in meq S L^{-1} of water. Multiply by 0.333 to get the value of SO_4 in the sample.

Sulphate (Jackson, 1973)

Apparatus/Reagents:

- **Standard Magnesium chloride ($MgCl_2$) solution (0.02 *N*)**: Dissolve 0.843 g of dried AR grade $MgCl_2$ in little excess of dilute HCl and dilute to 1 L with distilled water. Standardized against 0.02N EDTA solution.
- **Standard Barium chloride ($BaCl_2$) solution (0.02 *N*)**: Dissolve 2.4428 g of AR grade $BaCl_2.2H_2O$ in about 300 mL of distilled water and dilute to 1 L. Standardize against 0.2*N* Ca solution.
- **Standard Versene solution (0.01 *N*)**: Dissolve 4.0 g pure EDTA-disodium salt in about 200 mL of distilled water and dilute to 1 L. Standardize against 0.2*N* Ca solution.
- **Eriochrome black T indicator**: Dissolve 0.5 g of EBT dye and 4.5 g of hydroxylamine- hydrochloride (AR grade) in 100 mL of ethyl alcohol.
- **Standard HCl (0.2*N*)**: Dilute appropriate volume of concentrated HCl so as to get approximately 0.02*N* solution. Standardize it against standard alkali to exact normality.
- **A buffer solution of NH_4Cl and NH_4OH**: Mix 8.25 g ammonium chloride with 115 mL ammonium hydroxide in one liter volumetric flask and adjust the pH to 10 by drop wise adding NH_4OH and make up the volume to 1 L.

Procedure

1. Take 3 aliquots of sample each of 5 mL.
2. In the first aliquot determine (Ca+Mg) in meq L^{-1} as described previously.
3. In the second aliquot determine carbonate and bicarbonate as described previously.
4. In the third aliquot, add 0.02N HCl equivalent to total alkalinity (carbonate + bicarbonates) plus slightly more to completely neutralize the alkalinity. Shake the solution to destroy carbonates. Add 2 mL of reagent barium chloride and boil for few seconds. Cool and add 10 mL of reagent buffer solution and 5 drops of reagent Eriochrome black T indicator. Titrate the solution against standard versenate solution to the first end point (blue or green). Add 1 mL of reagent $MgCl_2$ and again titrate to obtain a second end point by titration with reagent EDTA.

Calculations

$$SO_4^{2-}\text{ (meq L}^{-1}) = (B + Ba + Mg - T) \times 1000/\text{ mL of aliquot taken} \quad (7.15)$$

Where, B is the amount of EDTA required for blank (distilled water), Ba and Mg are the meq of barium (Ba) and magnesium (Mg) added, T is the meq of EDTA added in the total titration of sample (including Ba and Mg).

Sulphate (Nephelometry)

Apparatus/Reagents: Nephelometric turbidity meter with sample cells. Alternatively a spectrophotometer, 420 nm with a light path of 2.5 to 10 cm, Magnetic stirrer, Timer with indication of seconds.

a. **Buffer solution A**: Dissolve 30g magnesium chloride, $MgCl_2.6H_2O$, 5g sodium acetate, $CH_3COONa.3H_2O$, 1g potassium nitrate, KNO_3, and 20 mL acetic acid CH_3COOH (99%) in 500 mL distilled water and make up to 1000 mL.

b. **Buffer solution B**: Only required if sample SO_4 concentration is less than 10 mg L^{-1}. Prepare as buffer solution A and add 0.111g sodium sulphate, Na_2SO_4.

c. **Barium chloride**, $BaCl_2$, crystals, 20 to 30 mesh

d. **Standard sulphate solution**: Dilute 10.4 mL standard 0.02N H_2SO_4 in to 100 mL (1.00 mL = 100μg SO_4)

e. **Standard sodium carbonate**, approximately 0.05 *N*. Dry 3 to 5g sodium carbonate, Na_2CO_3, at 250°C for 4 h and cool in a desiccator. Accurately weigh 2.5±0.2g to the nearest mg, dissolve in distilled water and make to 1 L.

f. **Standard H_2SO_4**, approximately 0.1*N*. Dilute 2.8 mL conc. sulphuric acid to 1 L. Standardize against 40.00 mL 0.05*N* Na_2CO_3 with about 60 mL distilled water, in a beaker by titrating potentiometrically to pH 5. Lift out electrodes, rinse into the same beaker and boil gently for 3 to 5 min under a watch glass cover. Cool to room temperature, rinse cover glass into beaker and finish titration to pH 4.3. Calculate normality of sulphuric acid:

   $$\text{Normality, } N = A\,B / (53.0 \times C)$$

   Here, A is Na_2CO_3 in grams weighed into the 1 L flask for preparing Na_2CO_3 standard (see e), B is mL Na_2CO_3 solution taken for standardization titration, C is the mL of acid used in standardization titration. In case potentiometric titration is not possible use bromocresol green indicator to complete the titration. The indicator is prepared by dissolving 100 mg bromocresol green sodium salt in 100 mL distilled water.

g. **Standard sulphuric acid**, 0.02*N*. Dilute the approximate 0.1*N* solution to 1 L. Calculate volume to be diluted as:

   mL volume = 20/N, such that N is the exact normality of the approximate 0.1*N* solution.

Procedure

1. Standardize nephelometer following manufacturer's instructions.
2. Measure the turbidity of sample-blank, a sample in which no $BaCl_2$ is added.
3. Measure 100 mL sample or a suitable portion made up to 100 mL, into a 250 mL conical flask. Add 20 mL buffer solution and mix. Continue stirring while adding spoonful of $BaCl_2$ crystals. Stir for about 60 s.
4. Measure turbidity of the sample at 5±0.5 min after stirring has ended.
5. Prepare SO_4 standards at 5 mg L^{-1} increments in the range of 0 to 40 mg L^{-1} SO_4 according to the following protocol:
6. Develop $BaSO_4$ turbidity for the standards as above and determine turbidity of the standards. Draw calibration curve between turbidity and SO_4 concentration, mg L^{-1}.
7. In case buffer solution B is used for samples containing less than 10 mg L^{-1} SO_4 run a reagent-blank with distilled water in place of sample. Develop turbidity for taking reading as before.

SO_4	5	10	20	30	40
Standard SO_4 solution	5	10	20	30	40
Distilled water, mL	95	90	80	70	60

Calculation

i. If buffer solution A is used, read SO_4 concentration for the sample from the calibration curve after subtracting the turbidity of sample-blank from the turbidity of the treated sample. If less than 100 mL sample was used, multiply the result by 100/mL sample volume.

ii. If buffer solution B is used, read SO_4 concentration in the treated sample from the calibration curve after subtracting the turbidity of sample-blank from the turbidity of the treated sample (same as in i). Subsequently, read SO_4 concentration for the reagents from the turbidity value of the reagent-blank from the calibration curve. Report the corrected SO_4 concentration in the sample after subtracting the reagent-blank SO_4 concentration from the sample SO_4 concentration.

Exchangeable Cations

Reagents for Preparation of Extracts

- **Alcoholic ammonium chloride, 1M:** Dissolve 53.5 g NH_4Cl in about 300 mL of water and add 600 mL of ethanol (95% v/v). Adjust the pH to 8.5-8.6 with concentrated ammonia solution and make up to 1 L. Store in a well stoppard bottle to prevent access of CO_2.

- **Ammonium ethanoate solution, 1M**: Dilute approximately 230 mL of glacial ethanoic (acetic) acid to 1L. Dilute approximately 220 mL of ammonium solution (approximately 35 % m/m NH_3) to 1 L. Mix the solutions together and adjust to pH 7.0 with ethanoic acid or ammonia. Dilute to 4 L.
- **Ethanol, 95%**
- **Potassium Chloride solution**: Dissolve approximately 100 g KCl in water and make up to 1 L.
- **Neutral 1 *N* NH_4OAC solution**: Dissolve 77.09 g of NH_4OAC in distilled water and make volume to 1 L. Adjust solution pH to 7.0 by adding acetic acid or NH_4OH solution.

Procedure

1. Weigh 5 g (± 0.01 g) of < 2 mm air dry soil into a 100 mL beaker. Add 20 mL of ammonium ethanoate solution. Stir and allow standing over night. Transfer the suspension to a filter funnel fitted with a Whatman no. 44 filter paper standing in a 250 mL volumetric flask. Thoroughly wash the beaker with ammonium ethanoate solution into the funnel. Leach the soil with successive 25 mL volume of ammonium ethanoate solution. Continue until 250 mL of filtrate is collected. Make up to the mark with ammonium ethanoate solution.
2. Use the extract so collected to determine potassium, calcium and magnesium.
3. Follow the same procedure with alcoholic ammonium chloride and use the leachate to determine sodium.
4. Sodium so determined minus the soluble sodium present in the soil saturation extract would give the exchangeable sodium.

Or

1. Take 5 g air dried soil in a 100 mL conical flask and add 25 mL of neutral normal NH_4OAC solution.
2. Shake on a mechanical shaker for 5 min and immediately filter through Whatman No. 1 filter paper. One can also shake through centrifugation (2500 rpm for 10 min) to obtain ammonium acetate extract. The extract can be diluted depending upon the expected ionic concentration.
3. Na and K are estimated by flame photometer as mentioned for saturated paste extract. Ca and Mg are also determined with the method explained previously.
4. The exchangeable cations are then determined by subtracting the soluble cation concentration from the values determined in this method.

Cation Exchange Capacity (CEC) (Richards, 1954)

Negatively charged colloidal (clay and humus) surface area attracts positively charged cations. The plant nutrients calcium, potassium, magnesium and cationic micro-nutrients are supplied to plants in large measure from exchangeable forms. The cation exchange sites immobilize cations but also keep them exchangeable and thus, make them available to plant roots. Cation exchange capacity is defined as the sum total of the exchangeable cations that a soil can adsorb. Cation exchange capacity can be determined by several methods, wherein all the adsorbed cations are replaced by a common ion, such as barium, potassium, sodium, or ammonium and then the amount of adsorbed barium, potassium, sodium, or ammonium is determined. The cation exchange capacity is expressed in terms of moles of positive charge per unit mass. For convenience, the cation exchange capacity is expressed in centimoles of positive charge per kilogram [cmol.(p^+) kg^{-1}] of soil or in meq 100 g^{-1} of soil.

Apparatus/Reagents: Funnel, Beaker, Filter paper, Flame photometer, Volumetric flask, Centrifuge tube.

- **Sodium acetate (1 *N*), Reagent A:** Dissolve 136 g of sodium acetate trihydrate in distilled water and dilute to a volume of 1 L (adjust the pH to 8.2 using dilute NaOH or CH_3COOH).
- **Ammonium acetate (1 N), Reagent B**: Add 57 mL of 99.5% acetic acid and 68 mL concentrated ammonium hydroxide, specific gravity 0.90, to about 800 mL water and dilute to 1 L (adjust the pH 7.0 using dilute NH_4OH or CH_3COOH). Optionally prepare NH_4OAc reagent salt (dissolve 77.08 g of ammonium acetate in distilled water and dilute to 1 L; (adjust the pH to 7.0).
- **Ethanol:** 60%
- **Standard sodium solution**: Dissolve 2.338 g of dry sodium chloride in 1 L distilled water. This will give a stock solution of 40 meq L^{-1} Na. Pipette 25 mL solution from the 40 meq L^{-1} stock solution of Na into a 250 mL volumetric flask, and make volume to capacity. This will give 4 meq L^{-1} Na solution. Pipette out 0, 5, 10, 15, 20, 25, 50 and 75 mL portions of this solution to 100 mL volumetric flasks and make the volume to 100 mL. These working standards are prepared by diluting the aliquot with 1*N* NH_4OAc solution. These will give 0, 0.2, 0.4, 0.6, 0.8, 1.2 and 3.0 meq L^{-1} Na concentrations. Calibrate the flame photometer for operation over the concentration range of 0 to 4 meq L^{-1}.

Procedure

1. Weigh 5 g soil, corrected for the air-dry moisture content, in a beaker and add about 33 mL of reagent A (preferably keep overnight to saturate the soil with Na)

2. Transfer the soil on to the funnel fitted with moist Whatman No. 42 filter paper. Add 33 mL reagent A on soil and leach it. When leaching is completed then add another 33 mL reagent A. Repeat one more washing with reagent A and discard the leachate.
3. Add 30 mL or so ethanol to soil so as to remove excess of reagent A. Continue washing with ethanol until the electrical conductivity of leachate from last washing is between 0.040-0.055 mS cm^{-1}. Discard the leachate of ethanol.
4. Replace the adsorbed sodium from the sample by extracting with three 33 mL portions of reagent B. Collect it in 100 mL volumetric flask. Make the volume of flask to 100 mL and determine the sodium concentration by flame photometer.

Alternative method

1. Weigh 5 g soil and place in a centrifuge tube. Add 33 mL reagent A, stopper the tube, and shake it for 5 minutes.
2. Remove stopper and centrifuge until the supernatant liquid is clear (usually 10 minutes at 2000 rpm). Decant the supernatant liquid as possible and discard. Repeat same procedure bringing the soil in centrifuge in suspension two more times, discarding the supernatant liquid each time.
3. After the last saturation, wash the stopper and use adsorbent paper to remove any acetate crystals remaining on the lip of the centrifuge tube.
4. Add about 30 mL ethanol to centrifuge tube, put stopper, shake for 5 minutes, remove stopper and centrifuge until the supernatant liquid is clear. Decant and discard the supernatant liquid. Continue washing with ethanol until the electrical conductivity of supernatant liquid from the last washing is between 0.040-0.055 mS cm^{-1}. Replace the adsorbed sodium from the sample by extracting with three 33 mL portions of reagent B. Make the volume 100 mL and determine the sodium concentration by flame photometer.

Calculations

$$\text{CEC [meq 100 g}^{-1}\text{ soil or cmol (P}^{+}\text{) kg}^{-1}] = \text{Na (meq L}^{-1}\text{) from standard curve} \times 10/\text{Weight of soil sample (g)} \quad (7.16)$$

Approximate method

The exchange capacity of the saline and sodic soils is dominated by Na^{+}, Ca^{2+} and Mg^{2+} with small amounts of K^{+}. If these are known, the CEC can be determined by adding the exchangeable cations.

$$\text{CEC [meq 100 g}^{-1}\text{ soil or cmol (P}^{+}\text{) kg}^{-1}] = \text{Ex Na}^{+} + \text{ExCa}^{2+} + \text{Ex Mg}^{2+} + \text{Ex K}^{+} \quad (7.17)$$

Here Ex denotes exchangeable form of the respective cation.

Gypsum Requirement

Theoretically, Gypsum Requirement (GR) can be calculated by the following relationship:

$$\text{GR (meq } 100\text{g}^{-1}) = \text{CEC (initial ESP – final ESP)}/100 \qquad (7.18)$$

Since 1 meq of gypsum 100 g^{-1} of soil is equivalent to 860 mg of gypsum and since 1 ha of soil to a depth of 20 cm may be taken to weigh 3,000,000 kg, the amount of gypsum theoretically required to treat this depth of soil is calculated as follows:

$$\text{GR (kg ha}^{-1}\text{ 20 cm}^{-1}) = \text{GR (meq } 100\text{g}^{-1}) \times 860\times10^{-6} \times 3\times 10^{6}$$

$$= \text{GR (meq}100\text{g}^{-1}) \times 2580 \text{ (kg)} \qquad (7.19)$$

If gypsum is not 100% pure, then the quantity of gypsum is determined by dividing the GR by purity of gypsum in fraction.

Schoonover (1952) method

Reagent: Prepare a nearly saturated gypsum solution of known calcium concentration. It is prepared by adding 5 gm of chemically pure $CaSO_4.2H_2O$ to one liter of distilled water. Shake vigorously for about 10 min on a mechanical shaker or 1 h by hand. Filter through whatman filter paper no. 1.

Procedure

1. Take 5 g of air dry soil in 250 mL conical flask.
2. Add 100 mL of the saturated gypsum solution. Firmly put a stopper and shake for 5 min on a mechanical shaker or 50 min by hand.
3. Filter the contents through whatman filter paper no.1. Transfer 5 mL of aliquot into 100 mL porcelain dish.
4. Determine the calcium and magnesium concentration as per the procedure outlined earlier in this chapter.
5. Run a blank using 5 mL of the saturated gypsum solution prepared for this determination.

   GR (meq 100 gm^{-1} soil) = 2 [Ca concentration in blank - Ca concentration in the filtrate]

Here Ca concentrations are in meq L^{-1}.

$$\text{GR for 15 cm soil layer (t ha}^{-1}) = \text{GR (meq 100 gm}^{-1}\text{ soil)} \times 1.72 \qquad (7.20)$$

Modified Schoonover method (Abrol *et al.*, 1975)

Gypsum requirement calculated using Schoonover method includes the gypsum required to neutralize the soluble sodium carbonate present in the alkali soils.

Abrol *et al.* (1975), therefore, suggested that the sample of the soil (5 gm) should be leached with 20 mL of 60% ethanol before it is used to determine gypsum requirement. Soluble sodium carbonate will be leached and the GR so calculated will be a better estimate than without leaching. Rest of the procedure continues to be the same as discussed in the previous paragraph.

GR using pH versus GR graph

The absolute quantity of exchangeable sodium percentage of the alkali soils is probably the most appropriate to determine GR. On the other hand, pH of the saturated soil paste (pH_s) or pH of 1:2 soil water suspension (pH_2) has been found to be well correlated with the ESP of the alkali soils of the Indo-Gangetic plains of India. Using this as a basis, Abrol *et al.*, (1973) prepared a graph between pH_2 and GR of the soils for three soil textures (Fig. 7.4). The use of this graph is a quick and fairly good method to estimate GR of alkali soils of the Indo-Gangetic plains. This method is particularly useful when skilled staff and laboratory facilities are not available to determine the exchangeable sodium and gypsum requirement.

Caution: The graph should not be used as a common practice for soils where relationship between pH and ESP is not established.

Once the GR is known, equivalence of other amendments can be determined using the equivalent factors in Table 4.3.

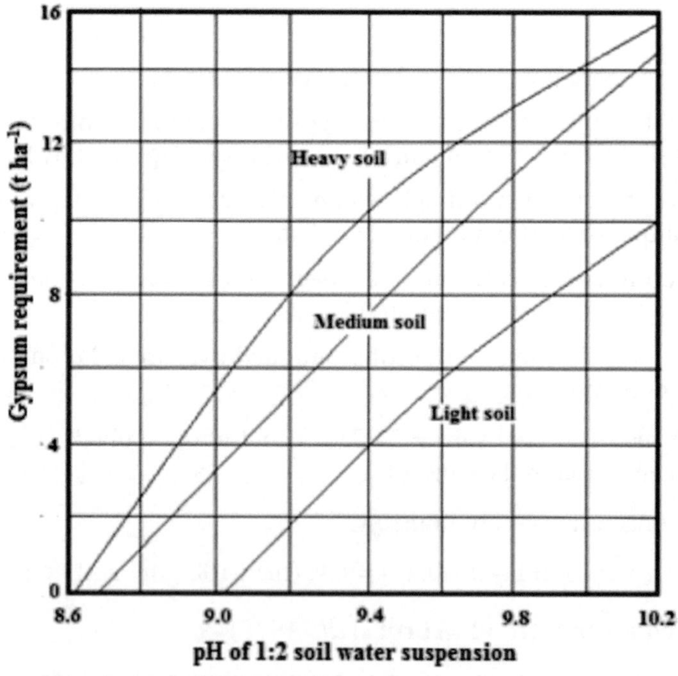

Fig. 7.4 pH and gypsum requirement of soils

Calcium Carbonate in Soil

Gravimetric loss of CO_2

Apparatus and Reagents: Pipette, Erlenmeyer flask, Hydrochloric acid 3 *N* (1:3)

Procedure

1. Pipette 10 mL of hydrochloric acid in Erlenmeyer flask, stopper with a cork and weigh (W1).
2. Weigh 1-10 gm of soil (W2) so that the soil contains about 0.1 to 0.3 g of calcium carbonate
3. Add soil to flask little at a time to prevent excessive frothing.
4. After the effervescence has subsided, replace the stopper loosely and swirl the flask. Continue swirling at frequent intervals and weigh after each swirling.
5. Once the weight does not change much in the 30 min. period, the reaction may be taken as complete.
6. Remove the stopper and swirl the flask for 10-20 s to remove any carbon dioxide that may still remain in the flask. Weigh the content with stopper on (W3).

$$\text{Carbon equivalent (\%)} = (\text{weight of } CO_2 \text{ lost} \times 227.4) / \text{weight of soil sample} \quad (7.21)$$

Weight of CO_2 lost = W1 +W2 – W3

Alternate method through Calcimeter (Allison and Moodie, 1965)

1. Take 1 g soil in a 100 mL conical flask
2. Take 20 mL HCl (1:3) in a 25 mL vial and carefully insert the same into the conical flask. After this, close the flask with the corked inlet tube of the calcimeter, which is installed in a chamber that is full of water.
3. The acid is carefully mixed with the soil sample and shaken properly for a few minutes. The volume of water displaced due to release of CO_2 is noted.
4. Similarly, volume of CO_2 released is noted by taking 0.1g $CaCO_3$ instead of 1g soil sample.

The $CaCO_3$ equivalent is then calculated as per the following relation:

$$CaCO_3 \text{ equivalent (\%)} = (\text{Weight of } CaCO_3 \text{ taken} \times \text{Volume of } CO_2 \text{ released for 1 g soil} \times 100) / (\text{Volume of } CO_2 \text{ released from the weight of } CaCO_3 \text{ taken}) \quad (7.22)$$

Calcium carbonate (Puri, 1930)

Apparatus and Reagents: Conical flask, Pipette, Burette, Volumetric flask, Hot plate

- **0.5N H_2SO_4**
- **0.1N $AlCl_3.6H_2O$**: Dissolve 8.05 g of aluminium chloride in a small quantity of water and dilute to 1 L with distilled water
- **Bromothymol blue (1% alcoholic solution)**: Dissolve 1 g of Bromothymol blue in 100 mL of alcohol
- **Bromocresol green (1% alcoholic solution)**: Dissolve 1 g of Bromocresol green in 100 mL of alcohol

Procedure

1. Weigh 10 g of soil and stir it with 100 mL of distilled water in a conical flask.
2. Add 0.2-0.5 g of $CaSO_4$ and bring the suspension to boiling and add 10 mL of 0.1*N* $AlCl_3.6H_2O$ solution.
3. Shake the suspension and add 10 drops of bromothymol blue and bromocresol green indicators and note down the color. If $CaCO_3$ is present (more than 1%), a green color develops. A golden yellow color indicates the absence of $CaCO_3$.
4. If carbonate is present, bring the contents of the flask to boiling point and titrate against 0.5*N* H_2SO_4. Boil for a minute or two after each addition and allow settling for a couple of minutes.
5. Continue the titration till golden yellow color persists on boiling for a couple of minutes. Note down the volume of 0.5*N* H_2SO_4 used for titration.

Since 1 mL of 0.5 *N* H_2SO_4 used is equal to 1 mL of 0.5 *N* $CaCO_3$ or 0.5 meq of $CaCO_3$ or 0.025 g $CaCO_3$

$$CaCO_3 \text{ in soil } (\%) = (0.025 \times Y \times 100) \; / \; \text{Weight of the soil} = 2.5\ Y/\text{weight of soil} \quad (7.23)$$

Here Y is the amount of 0.5*N* H_2SO_4 used.

Following guidelines on $CaCO_3$ and its impact on crops are then used to interpret the results

$CaCO_3$ (%)	Soil rating
< 5% $CaCO_3$	Fit for all crop
5-10% $CaCO_3$	Fit for all crops except citrus
>10% $CaCO_3$	Unfit for all crops

Inorganic Carbon

Inorganic carbon (%) is obtained by multiplying the $CaCO_3$ (%) by 0.012.

Oxidizable Organic Carbon

Wet digestion (Walkley-Black, 1934)

Apparatus and Reagents: Conical flask, Thermometer

- **Potassium dichromate (1*N* $K_2Cr_2O_7$) solution**: Dissolve 49.04 g of analytical grade $K_2Cr_2O_7$ (dried at 105°C for 2 hr) in distilled water and make the volume to 1 L.
- **Concentrated H_2SO_4** (36*N*), not less than 96% (specific gravity 1.84).
- **Ortho-phosphoric acid** (85%) or **sodium fluoride** (NaF).
- **Diphenylamine indicator**: Dissolve 0.5g diphenylamine indicator in a mixture of 100 mL concentrated sulphuric acid and 20 mL distilled water. Store in an amber colour bottle.
- **Ferrous ammonium sulphate [0.5*N* $FeSO_4(NH_4)_2SO_4.6H_2O$] solution** viz., **Mohr's salt**: Dissolve 392 g analytical grade ferrous ammonium sulphate (FAS) in distilled water, add 15 mL concentrated sulphuric acid and make volume to 2 L with distilled water.

Procedure

1. Take 2.0 g of soil (ground and sieved through 0.5 mm sieve) in a 250 mL conical flask.
2. Add 10 mL of 1*N* K_2 Cr_2O_7 solution through a pipette and shake gently to mix the contents.
3. Add 20 mL of concentrated sulphuric acid while swirling the flask slowly. Lot of heat of dilution is produced because the reaction is exothermic in nature.
4. Cool the contents to room temperature in dark. Note that the flask should be kept on a dry tile or asbestos sheet.
5. Add 0.5 g NaF or 5.0 mL of ortho-phosphoric acid, 100 mL of distilled water and 10 drops of diphenylamine indicator solution. Shake vigorously to mix.
6. Then titrate against 0.5*N* FAS solution till the colour changes from violet to bright bottle green through blue.
7. Note the volume of FAS solution used for titration.
8. Carry out a blank titration.

Note: While estimating organic carbon in saline soils using this method there will be some error as Cl^- ions in saline soils are also oxidized by dichromate ions. It will underestimate of the oxidizable organic carbon. Therefore, soils should first be treated with silver sulphate/mercuric sulphate/mercuric oxide associated with potassium dichromate and sulphuric acid to precipitate chlorides. For this purpose 1.25g of silver sulphate is dissolved in 100 mL of sulphuric acid and used for saline soils instead of only sulphuric acid.

Calculations

Total organic carbon in soil (%) = [N × (V_1- V_2)/ Weight of soil] × 0.39 (7.24)

Here, N is the normality of FAS solution, V_1 is the volume of FAS used for the sample (mL) and V_2 is the volume of FAS used for the blank (mL).

Since 1 mL of 1*N* $K_2Cr_2O_7$ = 3×10^{-3} g carbon, the factor 0.39 is given as $3\times10^{-3}\times 100 \times1.3$. The factor 1.3 is used with the assumption that there is 77% recovery since the oxidation of the organic matter in this procedure is incomplete.

Organic matter (%) = Total organic carbon (%) × 1.724 (7.25)

Here, 1.724 is the Van Bemmelen factor as explained below.

The organic carbon content in mineral surface soils range from 1.2 to 3.5%. Since soil organic matter averages about 58% of organic carbon, the organic matter in soils generally range from about 2 to 6% organic matter (% OM = % C × 1.724. The factor 1.724 is given by 100/58). A close relationship also exists between organic carbon and nitrogen in soils. Since most organic matter average about 5% nitrogen so that the N:C ratio is 1:11.6. Therefore by multiplying the soil organic matter percentage by 0.05 an approximate value of the soil nitrogen in percentage is obtained.

Soil organic matter by loss on ignition

Loss on ignition (LOI) method is generally used to determine the OM content of soils by weighing the sample before and after an ashing (Storer, 1984; Bisutti et al., 2004). The method is based on the principle that at the chosen temperature, all OC is converted to CO_2. As the soil organic matter is oxidized at a moderate to high temperature, with the weight loss being proportional to the amount of soil organic matter (SOM) in the sample (Konen et al., 2002).

Apparatus: Oven, Muffle furnace or oven-capable of maintaining a temperature of 500°C, Porcelain crucibles, Balance with accuracy of 0.001 g, Scoop - 1 cm^3 capacity

Procedure

1. Take the weight of empty crucibles (Wc).
2. Accurately weigh 2 g of the soil sample into the crucible, and record the weight of the crucible and soil sample (Wcs).
3. Dry the soil in the crucible in an oven set at 105°C for one hour.
4. Remove the crucible and allow cooling for 15 minutes.

5. Weigh the crucible with the dried sample (Wos).
6. Place the crucible into a muffle furnace equilibrated at 500°C for two hours.
7. Remove the crucibles and allow cooling for one hour.
8. Place the crucible and the ashed sample on the balance and record the weight (Wca).

Calculations

1. Moisture (%):

$$\text{Moisture (\%)} = 100\,(\text{Wcs} - \text{Wos})/(\text{Wcs} - \text{Wc}) \tag{7.26}$$

2. Loss on ignition (%)

$$\text{LOI (\%)} = 100\,(\text{Wos} - \text{Wca})/(\text{Wos} - \text{Wc}) \tag{7.27}$$

Here LOI is percent weight loss on ignition, Wc is the weight of empty crucible, Wcs is the weight of crucible plus soil before drying at 105°C, Wos is the weight of crucible plus soil after drying at 105°C and Wca is the weight of crucible plus ashed soil.

$$\text{Organic matter, OM (\%)} = 1.724 \times \text{LOI} \tag{7.28}$$

Total Carbon or Total Organic Carbon

Soil contains carbon in the form of organic and inorganic carbon. The sum of the two forms constitutes the total carbon of soil. Carbon/Hydrogen/Nitrogen/Sulphur (CHNS) analyzer is used to measure the total carbon while total organic carbon (TOC) analyzer is used to measure the total organic carbon of soil.

- Finely ground air dried soil sample (say 50-70 mg) is wrapped in tin boats (Supplied by the manufacturer), weighed and placed in the instrument.
- The direct values of total carbon and total nitrogen are obtained. Knowing the exact weight of the soil, results can then be expressed in %.
- Similar procedure is used to estimate total organic carbon of soil by using the instrument TOC analyzer.

Following guidelines are then used to rate the soil with respect to its $CaCO_3$ content

$CaCO_3$ (%)	Soil rating
< 5.0 g OC kg^{-1}	Low
5.0 to 7.5 g OC kg^{-1}	Medium
> 7.5 g OC kg^{-1}	High

Turbidity

Turbidity can be measured either by its effects on the transmission of light which is termed as turbidimetry or its effects on the scattering of light which termed as Nephelometry. Turbidimetry can be used for sample with moderate turbidity and Nephelometer for sample with low turbidity. Higher the intensity of scattered light higher the turbidity.

Apparatus and Reagents: Nephelometric turbidity meter with sample cells, Cuvettes (it take the samples for measurements)

- **Solution I.** Dissolve 1.000g hydrazine sulphate, $(NH_2)_2.H_2SO_4$ in distilled water and dilute to 100 mL in a volumetric flask.
- **Solution II.** Dissolve 10.00 g hexamethylenetetramine, $(CH_2)_6N_4$, in distilled water and dilute to 100 mL in a volumetric flask.
- **1000 NTU suspension**: In a 100 mL volumetric flask, mix 12.5 mL solution (I) and 12.5 mL solution (II) let them stand for 24 h at 25 ±3°C. Dilute to mark and mix. The turbidity of the suspension is 1000 NTU. The suspension is stable for up to 1 year. Dilute 1000 NTU stock solution with distilled water to prepare dilute standards just before use to be discarded after their use.

Procedure

1. Switch on the instrument and keep it on for some time. Calibrate nephelometer according to manufacturer's operating instructions as follows.
2. Select appropriate range depending upon the expected turbidity of the sample. Set zero of the instrument with turbidity free water using a blank solution and adjust 0 with zero knob.
3. In another cuvette take standard suspension diluted to say 200 NTU as standard. Take its measurements and set the display to the value of the standard suspension with the calibration knob. Run at least one standard in each range to be used.
4. Gently agitate sample. Dilute if necessary to be in the desired range for which instrument has been calibrated. Wait until air bubbles disappear and pour sample into Cuvette. Read turbidity directly from instrument display.

Calculation

$$\text{Turbidity} = A\,(B + C) / C$$

Here A is the NTU found in diluted sample, B is the volume of dilution water, C is the sample of volume taken for dilution

REPORTING OF THE RESULTS

To avoid changes in the physico-chemical properties of the samples at high temperatures, air dried samples are commonly used for analysis. On the other hand, it is customary to report the results on oven dried basis. Therefore, the results need to be multiplied by a moisture factor (F) that is calculated as per the following procedure.

Take exactly 10 g of air dry soil passed through 2 mm sieve. Keep the sample in an aluminium can and dry it at 105 ^{0}C for 24 h. Cool in a dessicator and weigh the can immediately to the nearest 1/100 of a gram. The F then can be given as

$$F = 10/\text{ wt of the oven dry soil (w)} = 10/w. \qquad (7.29)$$

Since w is less than 10, F is always more than 1.0. The analytical results are thus reported as

$$\text{Analytical result on oven dry basis} = \text{Analytical result on air dry basis} \times F \qquad (7.30)$$

CROSS-CHECKING CHEMICAL ANALYSIS

For consistency and reliability, it is always prudent and useful to check the chemical analysis of both soils and water. A general way to do so is to use the various relations that exist between the various parameters independently determined in the laboratory.

Electrical Conductivity and Total Cation Concentration

The first check is made by comparing the total cation and anion concentrations. The total soluble-anion and cation concentration or content expressed on an equivalent basis, should be nearly equal.

The EC of soil solutions and saturation extracts when expressed in dS m^{-1} at 25°C is approximately equal to the 1/10 of the total soluble-cation concentration in milli-equivalents per liter for solutions having EC < 5 dS m^{-1}.

Carbonate and Bicarbonate Concentrations and pH

Presence of carbonate ions is always associated with high pH. If carbonates are present in titratable quantities, the pH reading of the extract must exceed 9.

Calcium and Magnesium Concentrations and pH

The concentration of calcium and magnesium in a saturation extract are always low, if carbonate ions are present in titratable amounts or at high concentration of bicarbonate ions.

ESP and pH

A soil with ESP of more than 15 will have pH in excess of 8.5 at the saturation percentage

An increase in pH reading of 1.0 or more, as the moisture content is changed from a low to a high value, has been found useful in some areas for detecting saline and alkali soils.

PREPARATION OF ALKALI SOILS

To conduct experiments on alkali lands, many times it is necessary to artificially prepare an alkali soil. The following procedure is used to prepare an alkali soil of known pH or ESP.

- Estimate the amount of soil required for the experiment / study. Properly sieve and dry the soil collected from normal (non-saline, non-alkali) fields
- Find out the initial ESP, pH_2 and CEC of the soil
- Spread the soil on polythene sheets in a uniform layer, not thicker than 0.5 cm
- Calculate the amount of salt required to increase the pH_2 of the soil for pre-decided ESP level (Approximate relationship between ESP and pH_2 for alluvial soils of the Indo-Gangetic plains is given below).

ESP	Approx pH_2
20	9.1
25	9.2
35	9.4
40	9.5
45	9.6
50	9.7
65	9.8
75	9.9
85	10.0
100	10.6

Calculation of salt requirement

To prepare the soil of ESP 50 using medium textured alluvial soils of the Indo-Gangetic plains with a CEC of 12.

$$ESP = [Ex.\ Na\ (meq\ 100\ g^{-1}) \times 100] \times CEC$$

$$50 = Ex.\ Na \times 100 / 12$$

If Ex. Na is denoted by X, then

$$X = 50 \times 12/100 = 6.0\ meq\ Na$$

$$6 \times 23 = 138\ mg\ Na\ 100\ gm^{-1}\ soil\ or\ 1.38\ g\ Na\ 1000\ g^{-1}\ soil$$

Using $NaHCO_3$ (Equivalent weight 84)

23 gm of Na is available in = 84 gm $NaHCO_3$

1.38 gm of Na is available in = 84 × 1.38/23 = 5.04 gm kg^{-1} soil

- Dissolve the total amount of salt for the given quantity of soil in water
- Spray the salt solution on the soil with spray pump, till it becomes saturated
- Cover the soil with polythene sheet
- The remaining salt water can be sprayed on following days
- After spraying the total salt solution, the soil is kept moist for 15-20 days to allow the complete reaction of salts
- After the soil comes in condition, mix it thoroughly 4-5 times. After 4-6 weeks, the soil pH_2 may be estimated before using it for experiment.

PREPARATION OF SALINE SOIL

Saline soils are usually used in pot and micro-plot experiments. To develop a given salinity level, application of salts like NaCl, $CaCl_2$ and Na_2SO_4 dissolved in water in the ratio of 7:2:1 gives good results as it is the ratio in which these salts are found in soils under the semiarid conditions. Other composition of salts could be used depending upon the kind of experiments. In this case also, take dry, grounded and sieved (2 mm) soil. Fill the pots with known weight of the soil.

To calculate the salt required to prepare a soil with EC_e of 20 dS m^{-1}, It is proposed to calculate the salt requirement for 1 liter of 20 EC water that would be able to saturate about 2.5 kg of soil (assuming porosity of the soil as 0.4 by weight). The value would vary with the kind of soil.

Using the approximate relation: meq of cations = 10 EC, we need about 200 meq of cations. Dividing into the ratio 7:2:1, we have

NaCl 20 × 7 = 140 meq

$CaCl_2$ 20 × 2 = 40 meq

Na_2SO_4 20 ×1 = 20 meq

NaCl (equivalent weight 58.5) = 58.5 × 140 = 8190 mg

$CaCl_2$ (equivalent weight 55.5) = 55.5 × 40 = 2222 mg

Na_2SO_4 (equivalent weight 71.0) = 71 × 20 = 1440 mg

Dissolve NaCl and $CaCl_2$ in approximate half of the total water and Na_2SO_4 in remaining half of the water.

Test the EC of the water before using it to saturate the soil.

PREPARATION OF SALINE WATER

Saline water is often required to conduct laboratory and some pot and micro-plot experiments. These are also required to irrigate crops to conduct specialized tests. Following sections describe the methodology to prepare waters of various qualities. The method mainly uses the simplified relation: EC_{iw} = 10× salt concentration meq L^{-1}

Preparing Saline Water using Single Salt

Calculate the salt concentration in meq L^{-1} required to prepare a water of given EC. For example, to prepare a water of EC_{iw} = 5 dS m^{-1}, the salt concentration is 50 meq L^{-1} (Use the simplified relation).

- Weight of salt required = Equivalent weight of the salt × 50
- Dissolve any of the salts reported in Table 7.4 in 1 L of water
- Test the EC and adjust, if required

Table 7.4 Amount of salt required to prepare a water of 5 dS m^{-1}

Salt	Equivalent weight (mg)	Weight of salt required (mg)
NaCl	58.5	2925
$CaCl_2$	55.2	2760
$Na_2\ SO_4$	71.0	3550

Preparing Saline Water using More than One Salt

Saline water of EC_{iw} of 10 dS m^{-1} having Na:Ca ratio as 50:50 and Cl:SO_4 as 80:20 is to be prepared. For this purpose, a two way table is prepared

	Chloride	Sulphate	Total
Na	30	20	50
Ca	50	-	50
Total	80	20	100

Then we have

Salt	Concentration (meq L^{-1})	Equivalent weight (mg)	Weight of salt required (mg)
NaCl	30	58.5	1755
Na_2SO_4	20	71.0	1420
$CaCl_2$	50	55.5	2775

- Weigh the salts as per the calculated weight in column 4 and dissolve in 1 L distilled water.
- Check EC and adjust, if necessary

Saline Water with desired EC and SAR

It is desired to prepare water whose EC_{iw} is 10 dS m^{-1} and SAR is 10. Once again a two way table is prepared. Let us assume that only chloride salts are used. Following two way table is prepared.

	Concentration (meq L^{-1})	Total
Na	50	50
Ca	25	25
Mg	25	25

Then we have

Salt	Concentration (meq L^{-1})	Equivalent weight (mg)	Weight of salt required (mg)
NaCl	50	58.5	2925
$CaCl_2$	25	55.2	1390
$MgCl_2.6H_2O$	25	71.0	2540

- Weigh the salts as per the calculated weight in column 4 and dissolve in 1 L distilled water.
- Check EC and SAR and adjust, if necessary

Saline Water with Desired EC and RSC

Let us say, that we need to prepare a water having an EC_{iw} of 2 dS m^{-1} and RSC of 5 meq L^{-1}.

Since RSC is given be the relation $(CO_3 + HCO_3) - (Ca + Mg)$, we can have both carbonates and bicarbonates besides chloride and sulphate. A two way table is prepared.

	CO_3	HCO_3	Cl	Total
Na	2	8	5	15
Ca	-	-	2	2
Mg	-	-	3	3
Total	2	8	10	20

Salt	Concentration (meq L^{-1})	Equivalent weight (mg)	Weight of salt required (mg)
NaCl	5	58.5	292
$CaCl_2$	2	55.2	111
$MgCl_2$.6H2O	3	101.7	305
Na_2CO_3	2	53.0	106
$NaHCO_3$	8	84.01	672

- Weigh the salts as per the calculated weight in column 4 and dissolve in 1 L distilled water.
- Check EC and RSC and adjust, if necessary

■■■■

Chapter - 8

METHODS OF DETERMINATION: PHYSICAL PROPERTIES

There are numerous soil physical parameters but only limited numbers are measured depending on the objectives. For the salt affected soils measurements generally include bulk density, particle density, porosity, soil water content at various suctions, particle size distribution, infiltration rate, hydraulic conductivity, aggregate stability, and dispersion etc. Measurement procedures of these properties are described in this chapter.

BULK DENSITY OF SOIL

Soil bulk density (ρb) is a measure of the oven-dry mass per unit volume of bulk soil. It depends on several factors like texture, structure, organic matter content, state of compaction etc. The bulk density of soil divided by the density of water is called the apparent specific gravity. Typically, sands pack more closely and values range from 1.4 to 1.9 Mg m^{-3}. Clays tend to bridge and cannot pack as tightly, giving values of 0.9 to 1.4 Mg m^{-3}.

Apparatus/Equipment: Core sampler, Metal cores, Aluminium boxes, Oven, Balance, Knife/spatula, Vernier caliper.

Core Sampler

A core sampler/cutter is usually made up of galvanized iron cylinder of 5.5–7 cm diameter and 15 cm height. The core cutter holds an assembly of a sectional brass cylinder of 5 cm diameter and 6 cm height with 2 brass rings of same diameter but 2 cm height on either side of it. These are screwed to the collar of sampler (2.5 cm high) from the upper side. The collar is attached to a thick hollow iron rod of nearly 85 cm in length. The other end of the rod is connected to a handle with a thick hollow iron piece in the middle, over which a hammer can be dropped. Simple soil cores manufactured locally can also be used to measure the bulk density.

Procedure

1. Measure the length (L) and diameter (2r) of the core (volume = $\pi r^2 L$) using vernier caliper and arrange them before going to the field.
2. In the field, place the core inside the core sampler. For taking samples, the assembled core cutter is positioned over a clean levelled surface

and pressed inside the soil by rotating the handles or by dropping a hammer over the central portion of the upper end of the thick rod until the edge of the collar comes to rest over the soil surface.

3. Loosen the grip of core cutter from soil around it by moving the stem forward and backward. Pull out the core cutter slowly from the soil and place it horizontally on surface. Unscrew the collar and carefully take out the rings and core containing soil.
4. Trim off the extra soil from both ends of the core with a sharp knife or spatula.
5. If it is proposed to weigh in the laboratory, place cores containing soil samples into strong plastic bags that can be sealed tightly with minimal air space. If weighing is done *in-situ*, transfer the soil core from the sampler to a pre-weighed aluminium can and take the weight. Later on put the same in the oven at 105°C for 24-48 hr till constant weight is achieved.
6. Take the dry weight of the soil with can.
7. Take out the sample and weigh the empty can.

Note: If the soil contains significant amount of gravel, weight of gravels may be taken off the final (dry) soil weight.

Observation Sheet

1. Soil sample identification number
2. Depth of sample (cm)
3. Inner diameter of the core (cm) = d
4. Length of the core (cm) = L
5. Sample volume, $V = (\pi \times d^2 \times L)/4$
6. Wt. of can (g) = W1g
7. Wt. of moist soil +can (g) = W2 g
8. Wt. of can + dry soil (g) = W3 g
9. Wt. of moist sample = W2 – W1 g
10. Wt. of dry sample = W3 – W1g
11. Wet bulk density = (W2-W1)/ V g cm^{-3} or Mg m^{-3}
12. Dry bulk density, ρb = (W3 – W1)/V g.cm^{-3} or Mg m^{-3}

Limitations: This method has few limitations. It cannot be used to determine the bulk density of sandy soil or wet soil. The core in such soils will not be able to retain the soil. Similarly, even in gravelly soil this method may not be practicable as to push the core may pose problems.

PARTICLE DENSITY OF SOIL

Soil particle density (ρs) is an important soil property as it appears along with the bulk density in the expression to calculate soil porosity. It also gives a fair view of the relative amount of organic matter versus the mineral particles in the soil sample. It is a measure of the oven-dry mass per unit volume of soil solids only. Each individual soil mineral has a characteristic particle density. Quartz, a predominant soil mineral, has a value of about 2.65 Mg m^{-3}, which is why this value is frequently taken to represent all soils. Organic matter (OM) (which is also a soil solid) can influence particle density. Soils with high OM (Mass of OM is much less per unit volume than soil minerals) may have much lower average particle density than soils similar in texture that are low in OM. Standard procedure calls for removal of OM before particle density analysis of such soils. However, this may produce significant error while calculating porosity of soils using ρs data. Particle density of most mineral soils is in the range of 2.60 – 2.75 Mg m^{-3} (representing particle density of quartz, feldspar and colloidal silicates which dominate the mineral fractions of soils).

Apparatus/Equipment: Pycnometer, Analytical balance, Beaker 250 mL, Filter paper, Air free distilled water, Funnel, Clean and dry cloth.

Procedure

1. Weigh a dry, clean pycnometer with glass stopper to the nearest 0.01 g, W1.
2. Transfer carefully 10 g of oven-dry soil to the dry pycnometer, replace the stopper and record the weight, W2. Mass of the soil Ws = W2- W1
3. Add air-free boiled and cooled water to the pycnometer until it is about 2/3rd full and allow it undisturbed for few minutes. Remove the entrapped air by gentle boiling for several minutes while gentle shaking or stirring lightly with a glass rod
4. Fill the pycnometer with air-free boiled and cool water to the brim and put stopper tightly so that extra water shoots out the hole in the stopper.
5. Carefully dry the outside of the pycnometer by wiping with a dry cloth and weigh to the nearest 0.01 g (W3).
6. Wash the pycnometer bottle and fill the pycnometer with air-free boiled and cool water to the brim and put the stopper tightly so that extra water shoots out of the hole in the stopper. Record the weight of pyconmeter filled with water (W4).

Particle density (ρs) is calculated by the following formula.

$$\rho s = [Ws] / [W4 + Ws - W3] \times \rho_{wt} \ \text{g cm}^{-3} \quad (8.1)$$

Here, ρ_{wt} is the density of water at water temperature t (g cm^{-3})

Porosity

The porosity, volume of voids divided by the total volume of the soil or the fraction of the soil space occupied by the air and water is often denoted by η. It is given as

$$\eta = (\rho s - \rho b)/\rho s = 1 - \rho b/\rho s \tag{8.2}$$

Here ρs and ρb are the particle and bulk density of the soil. The porosity in percentage is calculated by multiplying the result by 100.

SOIL MOISTURE CONTENT

As water is the most limiting factor in the arid and semi-arid regions, soil moisture determination is of major significance. Soil moisture influences crop growth by altering the matric stress to the plants. Besides, it affects nutrient availability and also influences nutrient transformations and soil biological behavior. Following 3 moisture contents are commonly measured.

- Moisture content of air dried soil also called hygroscopic moisture content
- Moisture content at saturation
- Moisture content at field capacity, the upper limit of available moisture content

Apparatus: Electric oven with thermostat, Desiccator, Aluminium cans with lids, Weighing balance

Moisture Content of Air Dry Soil

Procedure

1. Weigh the metal can with lid, say W_0
2. Pour air-dry soil (< 2 mm) into a previously dried (105°C) and weighed metal can with lid and weigh, say W_w. Find out the weight of the soil, $W1 = W_w - W_0$.
3. Dry in an oven, with the lid unfitted, at 105°C overnight (normally for 24-48 hr).
4. Cool in a desiccator for at least 30 min, fit the lid and re-weigh, say W_d. Find out the weight of the dry soil, $W2 = W_d - W_0$.

$$\text{Moisture content (\%)} = 100\,(W1 - W2)/W2$$

Saturation Percentage

Procedure

One can use the soil prepared for drawing the saturation extract of the soil (Chapter 7) or prepare the saturated soil as per the procedure outlined in that section.

1. Weigh the metal can with lid, say W_0
2. Transfer a portion of the saturated soil paste into the can and weigh. Determine the weight of the moist soil, say W_w. Find out the weight of the soil, $W1 = W_w - W_0$.
3. Dry in an oven, with the lid unfitted, at 105°C overnight (normally for 24-48 hr).
4. Cool in a desiccator for at least 30 minutes, fit the lid and re-weigh. Find out the weight of the dry soil, say W_d. Find out the weight of the dry soil, $W2 = W_d - W_0$.

$$\text{Saturation moisture content (\%)} = 100\,(W1 - W2)/W2 \tag{8.3}$$

Field Capacity (Field Method)

Apparatus: Polythene sheets, Spade, Soil auger, Moisture boxes/cans, Balance, Oven

Procedure

1. Select a level and uniform plot having dimensions of about 5 m × 5 m. Remove any loose material from the surface (weeds, pebbles, etc.). Make bunds all around the plot.
2. Fill sufficient water in the plot so that soil gets fully saturated.
3. Cover the plot area with a polythene sheet so as to check evaporation.
4. Beginning about 24 h after the sheet has been put on the soil surface, take soil sample from the centre of the plot from the desired layer, and determine soil moisture content daily as previously discussed in this section till the values of successive days are nearly equal. Field capacity is normally attained in the field after 24 to 72 hours of saturation depending upon the soil texture.
5. The constant moisture content so determined is the field capacity moisture content. One can also plot the daily readings on a graph paper. The nearly constant lowest reading is taken as the value of field capacity of the soil.

Field capacity (Laboratory Method, Bandyopadhyay et al., 2012)

Procedure

1. Use one of the 15-cm cores collected during the measurement of bulk density. Carefully remove the caps so that soil does not fall out of the core. Put a double layer of the cheese cloth on the bottom of the core with a rubber band. Separately weigh the cheese cloth and rubber bands before putting them on the core (Wc).
2. Weigh the core with the cheese cloth and rubber bands (W1).

3. Set the core in a pan of water until it is almost immersed; leave it in the pan to soak up water for 24 hours until it is completely saturated. Add extra water to the pan if needed
4. Remove the core from the pan, and let excess water drain until it stops dripping.
5. Weigh the core with soil at saturation point (W2). Put the core on a grate so that water can drain out of it; cover the top of the core with some plastic wrap to prevent evaporation. Let it drain for 24-48 hours.
6. Weigh the core with soil at field capacity after 24 (W3), 48 hours (W4) and 72 hours (W5)
7. Carefully remove the cheese cloth and replace with an aluminium foil; put core in the drying oven for 24 -48 hr at 105^{0}C. Weigh the wet piece of cheese cloth and rubber bands used on the core (W6); its weight needs to be subtracted from the total weight of the core + soil to get the actual weight of the water in the soil.
8. Remove the core from the oven, cover top with plastic wrap to prevent it sucking moisture from the atmosphere. Cool and weigh the dry core after removing the plastic wrap and aluminium foil (W7).
9. Calculate the amount of water in the soil at saturation and field capacity after 24, 48 and 72 hours by subtracting the weight of the core + dry soil from the weight of the core + saturated soil, respectively:
10. Weigh the empty core, W8

Weight of the soil + core = W1 – Wc = Wsc

Weight of the soil + core + water (saturation) = W2 –W6

Weight of the soil + core +water (24 hrs drain) = W3 –W6

Weight of the soil + core +water (48 hrs drain) = W4 –W6

Weight of the soil + core +water (72 hrs drain) = W5 –W6

Weight of the soil (dry)+ core = W7

Weight of the soil (dry) = W7 –W8

Moisture content at soil saturation (% dry weight) =
100 x (W2-W6 –W7)/(W7-W8)

Moisture content at 24 hours (% dry weight) =
100 x (W3-W6 –W7)/(W7-W8)

Moisture content at 48 hours (% dry weight) =
100 x (W4-W6 –W7)/(W7-W8)

Moisture content at 72 hours (% dry weight) =
100 x (W5-W6 –W7)/(W7-W8)

The constant weight after 24 to 72 hours or after 72 hours may be taken as the field capacity of the soil.

PARTICLE SIZE DISTRIBUTION

Particle size distribution is an important parameter in soil classification and has implications for available soil water, soil aeration, and nutrient availability to plants. Soil texture is also an important property of the agricultural soils, as it influences fertility, drainage, water-holding capacity, aeration, tillage, and soil strength. Because texture is so important, this test is always performed as a standard physical soil test. Laboratory procedures normally estimate percentage of sand (0.05 – 2.0 mm), silt (0.002 – 0.05 mm), and clay (<0.002 mm) fractions in soils. The two common procedures used for particle size analysis are the pipette gravimetric method (Piper, 1942; Chopra and Kanwar, 1991) or the hydrometer method (Ashworth et al., 2001).

Stokes Law

The hydrometer method of silt and clay measurement relies on differential settling velocities of different sizes of particle in a water column. They are separated by a process known as sedimentation, which is based upon the fact that larger particles fall more quickly through water than smaller ones. If all other factors are constant, then the settling velocity is proportional to the square of the radius of the particle (Stoke's Law).

$$V = 2\,r^2\,(\rho s - \rho_{dl})\,g\,/2\eta \qquad (8.4)$$

Here V is the velocity of fall (cm s^{-1}), r is the equivalent radius of the particle (cm), ρs is the density of particle (g cm^{-3}), 2.65 is a (usual) good approximation, ρ_{dl} is the density of liquid (g cm^{-3}), for water this is ~1.0, g is the acceleration of gravity (cm s^{-2}), value at the sea level is 981 and η is the viscosity of liquid (g $cm^{-1}sec^{-1}$). In practice, the density of water and its viscosity are both affected by temperature. Soil particles fall rapidly with increasing temperatures. Therefore, it is a usual practice to measure and make corrections for the temperature of the liquid.

Apparatus and Reagents: Soil dispersing stirrer, Hydrometer with Bouyoucos scale in g L^{-1}, Pipette, Stop watch, Hot plate, Drying oven, Weighing balance, accurate to 0.5 g, Steam plate or boiling water bath, Desiccator, Beakers, Measuring cylinders- 1000 mL, The sieve set of 10 and 70 mesh, Whatman No. 50 filter paper litmus paper, stirrer

- **Hydrogen peroxide**: 6% H_2O_2
- **HCL**: 2*N* HCl
- **Dispersing solution**: $NaHSO_3$, 1*N* sodium hydroxide or sodium hexametaphosphate
- **Amyl alcohol**

Pipette Gravimetric Method

Removal of organic matter

1. To remove organic matter, place about 20 g of oven dry soil that has been passed through a 2 mm sieve in a 400 mL beaker
2. Add 20 mL of 6% H_2O_2 solution (30 to 40 mL of H_2O_2 if the sample contains more than 3.5% organic matter) and allow to stand overnight. Add another 10 mL of H_2O_2 solution and cover the beaker with a watch glass.
3. Place the beaker on a steam plate or over boiling water bath for 15 minutes. Stir when necessary to control frothing. Repeat the heat treatment with another 10 mL of H_2O_2 solution if the colour of the soil has not become light enough until disappearance of frothing to indicate that the organic matter has been destroyed. Boil the contents gently for another 5 minutes.
4. If manganese compounds are present in the sample (manganese compounds are indicated if H_2O_2 fails to destroy the organic matter), begin by placing the soil in 40 mL of water and add 0.5 g of $NaHSO_3$. Boil until water is half gone, and then continue as above.

Removal of carbonates

1. After complete removal of the organic matter, cool the beaker and clean the sides of the beaker with a rubber pestle. Add 25 mL of 2*N* HCl and stir well (add more acid at the rate of 2.5 mL for each percent of $CaCO_3$ present above 2%. Stir occasionally while the reaction proceeds for about one hour. Test after one hour, to make sure that an excess of acid is present. Filter the contents through a Buckner funnel using Whatman filter paper No. 50.
2. Wash the soil with hot distilled water till the excess acid is removed (This can be tested with a litmus paper or putting a drop of silver nitrate solution in the filtrate).
3. After the soil has been washed nearly free from acid and drained, transfer it from the Buckner funnel to 1000 mL long neck beaker or in a dispersing cup. After removing most of the soil, spread the filter paper on clock glass and clean the traces of soil from it by distilled water jet. When all the soil has been removed, roll the filter paper into a loose ball, wet it and thoroughly squeeze like a sponge. Repeat this process until no more turbid liquid can be obtained. Also wash the Buckner funnel rinsing it with jet of distilled water.
4. When the whole of the soil has been transferred to the 1000 mL beaker, add 10 mL of 1*N* sodium hydroxide or sodium hexameta-phospahte.

5. Place the beaker beneath the dispersing unit, lower the propeller into position and set the later into motion, controlling the speed. Adjust the speed so that vigorous stirring is affected, but there is no splashing. Continue for 10-15 min, then raise the motor propeller and rinse the propeller blades into the beaker.
6. Place a 50 μm (70 mesh) sieve, held over the mouth of a funnel which is placed over a 1000 mL cylinder. Pour the suspension on to the sieve and with a stream of distilled water from the wash bottle, wash as much material as possible through the sieve. Transfer the coarse material thus left on the sieve to a weighed dish. Dry it at 105°C in an oven and find its weight. Calculate the percentage of coarse sand as follows :

   Weight of dish $= W_1$ g

   Weigh of dish + dry coarse sand $= W_2$ g

   Weight of coarse sand $= W_2 - W_1$ g

   Weight of the soil sample taken $= W_0$ g

   % coarse sand $= 100 (W_2 - W_1) / W_0$ (8.5)

7. Fill the cylinder to the 1000 ml mark with distilled water. Check the temperature of the suspension. Stopper the top of the cylinder and thoroughly mix the contents by inverting it several times while shaking.
8. Set the cylinder on the table; record the time (hour, minute, second) immediately.
9. At the determined time against the suspension temperature (Table 8.1) insert a pipette into the suspension to a depth of 10 cm and withdraw a 25 mL sample. (Mark the depth on the pipette with a grease pencil)

Table 8.1 The time of sedimentation for clay and silt fractions at different temperature at 10 cm depth

Temperature	Time for clay		Time for silt + clay	
^{0}C	Hrs.	Min.	Min.	Second
8	11	-	6	40
9	10	40	6	30
10	10	25	6	20
11	10	10	6	10
12	9	54	6	10
13	9	35	5	50
14	9	20	5	40
15	9	05	5	30

Temperature	Time for clay		Time for silt + clay	
^{0}C	Hrs.	Min.	Min.	Second
16	8	50	5	20
17	8	35	5	10
18	8	25	5	05
19	8	10	5	00
20	8	00	4	48
21	7	50	4	40
22	7	40	4	30
23	7	25	4	30
24	7	15	4	20
25	7	05	4	15
26	6	55	4	10
27	6	45	4	05
28	6	40	4	00
29	6	30	3	55
30	6	20	3	50
31	6	15	3	45
32	6	05	3	40
33	5	55	3	35

10. Place the sample in a dish/beaker that has been weighed accurately to 0.01 g. Evaporate the water by placing the dish in an oven set at 105°-110°C for at least 24 hours. Weigh the sample to 0.01 g. The % silt + clay in the sample is calculated as follows:

    Weight of dish = W_1 g

    Weigh of dish + silt + clay = W_2 g

    Weight of silt +clay = $W_2 - W_1$ g

    $$\% \text{ silt+clay} = (W_2 - W_1) \times 1000 \times 100/ (25 \times W_0) = Y \quad (8.6)$$

11. Shake the contents of the cylinder again for 1 min and record the temperature and time. Take the sample in the same way at the depth of 10 cm after the settling time for clay (Table 8.1). Dry it and weigh and calculate the clay percentage as follows:

    Weight of dish = W_1 g

    Weight of dish + clay = W_2 g

    Weight of clay = $W_2 - W_1$ g

    $$\% \text{ of clay} = (W_2 - W_1) \times 1000 \times 100/ (25 \times W_0) = X \quad (8.7)$$

% silt = Y - X (8.8)

Fine sand

It can be calculated either by difference or by sedimentation.

% Fine sand = 100 – (Silt + clay) (8.9)

Sedimentation method

Pour away the bulk of the supernatant liquid and transfer the sediment of cylinder to a 500 mL beaker, make up with water to a height of 10 cm above the base which should be marked with a pencil using foot rod. Stir well and allow standing for the requisite period taken from the silt column in the table (Table 8.1) against the observed temperature. Then pour away the turbid suspension. Fill beaker again to the mark with distilled water and repeat this process till the liquid is no longer turbid at the end of the period. The residue is the fine sand which is dried at 105°C and weighed. Calculate the percentage of fine sand as follows:

Weight of dish = W_1 g

Weight of dish + fine sand = W_2 g

Weight of fine sand = W2 – W1 g

Weight of initial soil taken = W_0 g

% of fine sand = $(W_2 - W_1) \times 100/W_0$ (8.10)

Cautions

1. If air dry soil is taken for analysis, then moisture correction should be applied.
2. During the drying process, some loss due to ignition may occur, the correction should be applied. Loss means the loss due to the decomposition of $CaCO_3$ and oxidation of organic matter.

Hydrometer Method

Follow the same procedure as in the case of pipette method and transfer suspension quantitatively into a 1000 mL calibrated cylinder (hydrometer jar), and bring to volume with water.

A. Determination of Blank

1. Dilute 60 ml dispersing solution to 1000 mL hydrometer jar with water.
2. Mix well, and insert hydrometer, and take hydrometer reading, Rb.
3. Take the temperature.

B. Determination of Silt plus Clay

1. Mix suspension in the hydrometer jar, using a special paddle or plunger carefully, withdraw the paddle, and immediately insert the hydrometer.
2. Disperse any froth, if needed, with one drop of amyl alcohol, and take hydrometer reading 40 seconds after withdrawing the paddle. This gives reading, Rsc. Also measure the temperature.

Adjusting the reading for blank and temperature

For every 1°C below 20°C, subtract 0.36 from the reading.

For every 1°C above 20°C, add 0.36 to the reading.

For example, if the blank reading is 3.5, and the 40 s reading is 22 at 21°C, then the corrected reading is given as:

$$\text{Corrected reading} = \text{Reading} - \text{blank} + (1\times0.36) = 22\text{-}3.5+[1\times0.36] = 18.14 \quad (8.11)$$

C. Determination of Clay

1. Mix suspension in the hydrometer jar with paddle; withdraw the paddle, leave the suspension undisturbed.
2. After 2 h, insert the hydrometer, and take hydrometer reading, Rc.

D. Determination of Sand

1. After taking readings required for clay and silt, pour suspension quantitatively through a 50 μm sieve. Wash sieve until water passing the sieve is clear.
2. Transfer the sand quantitatively from sieve to a 50 mL beaker of known weight.
3. Allow the sand in the beaker to settle, and decant excess water. Dry the beaker with sand at 105°C overnight. Cool in a desiccator, and re-weigh beaker with sand. Find out the weight of the dry sand.

$$\text{Silt+Clay (\%)} = (\text{Corrected Rsc}) \times (100/\text{Oven dry soil, g}) \quad (8.12)$$

$$\text{Clay (\%)} = (\text{Corrected Rc}) \times (100/\text{Oven dry soil, g}) \quad (8.13)$$

$$\text{Sand (\%)} = \text{Sand weight, g} \times (100/\ \text{Oven dry soil, g}) \quad (8.14)$$

Oven dry soil is the corrected oven dried initial weight of the soil taken. Silt is determined from silt + clay by subtracting clay %. Sum of silt and clay + sand should be 100%. The magnitude of deviation from 100 is an indication of the degree of accuracy.

Remarks: If possible, all hydrometer jars should be placed in a water bath at constant temperature of 20°C to avoid the need for temperature corrections.

Expression of results

Tabulate the analysis showing the oven dry percentage of each fraction in the original soil as follows:

1. Coarse sand, 2. Fine sand, 3. Silt (ignited), 4. Clay (ignited), 5. Loss on solution or loss on ignition + $CaCO_3$, 6. Moisture (if air dry soil is taken) 7. Total.

Note: The sum should approximate to 100 and if the total is less than 98.5 or more than 102, the analysis should be repeated.

After the determination of sand (coarse + fine sand), silt and clay percentage, the soil texture is determined from the soil textural triangle (Fig. 8.1).

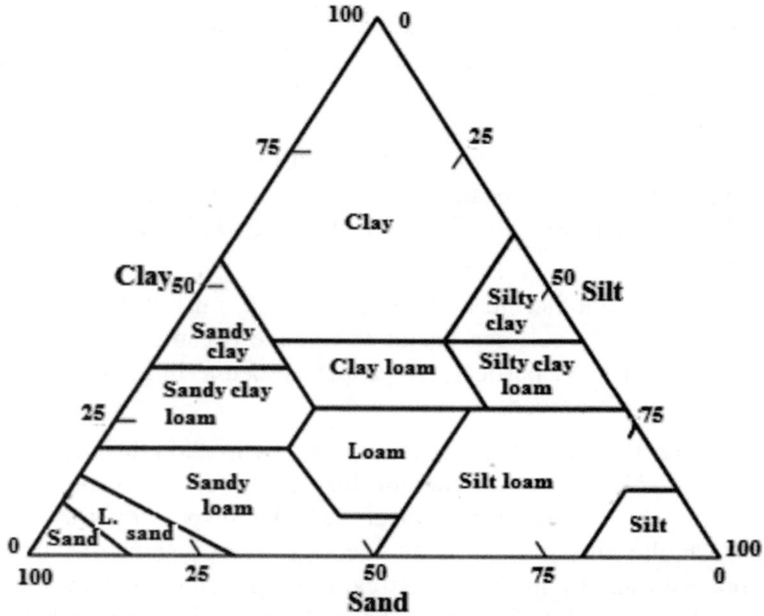

Fig. 8.1 The textural triangle for assigning the soil texture (after Hillel, 1980)

SOIL AGGREGATE STABILITY

Water Stable Aggregates (Laboratory Method)

The aggregate stability to water is determined by the principle that unstable aggregates break down easily than stable aggregates. Yoder (1936) first developed wet-sieving techniques using a nest of six sieves secured to a holder, which moves up and down in standing water in a drum through a distance of 3.18 cm at a rate of 30 cycles min^{-1} for 30 min. Later on Yoder's apparatus has been modified in several ways to suite specific purposes.

Apparatus and Reagents: A wet sieve shaker, Set of sieves with sieve openings of 5.0, 2.0, 1.0, 0.5, 0.25 and 0.1 mm diameter stacked in descending order, H_2O_2, Dispersing agent.

Procedure

1. Take 50 g (M0) of air-dried soil aggregates (passed through 8 mm and retained on 5 mm sieves). Place the aggregate sample gently onto the top of the nest of sieves.
2. Slowly bring the level of the water in the tank from where water just begins to wet up the aggregates. Allow the aggregates to wet for 10 minutes. Switch on the mechanical oscillator to move the nest of sieves up and down with a frequency of 30 cycles per minute and a stroke of 3 cm. Continue for 10 min.
3. Remove the nest of sieves from the water and allow it to drain for some time. Gently separate the sieves.
4. Pour and wash out the contents of each sieve into a pre-weighed beaker and let them settle for 24 h. Then carefully decant off excess water making sure not to lose soil. Dry the beakers along with the contents in an oven at 105°C for 24-48 hr. Weigh the beaker along with the dry soil aggregates and find out the weight of the dry soil aggregates in each beaker. Note down the weights as M1, M2, M3, M4, M5 and M6 for 5.0, 2.0, 1.0, 0.5, 0.25 and 0.1 mm sieves respectively.
5. Carry out another test. Now add 30-40 mL of H_2O_2 into beaker containing dry soil aggregates to remove organic matter, the binding agent of primary particles. Repeat the process till the organic matter is completely removed.
6. Transfer the contents into the dispersion cup. Add dispersing agent (10 mL of 1*N* NaOH) and enough distilled water to fill the cup for easy stirring by a mechanical stirrer. Stir the suspension for 10 minutes. Wash the suspension on an identical set of sieves as used in the previous test by means of a stream of tap water. Transfer the contents on the sieves into beakers. The primary particles in each beaker are oven-dried and weighed. Note down the weights as M7, M8, M9, M10, M11 and M12 for 5.0, 2.0, 1.0, 0.5, 0.25 and 0.1 mm sieves respectively.
7. Oven dry and weigh the soil passing through 0.1 mm sieve (M13) received in pan.

To calculate the percent water stable aggregate, use the following equation.

$$\text{Total aggregation (\%)} > 0.25\text{ mm} = [(M1+M2+M3+M4+M5)/M0] \times 100 \qquad (8.15)$$

Total aggregation (%) > 0.25 mm corrected for primary particles = 100 $[\sum(M_i - M_{i+6})]/M_0$

Such that i varies from 1 to 5, the number of sieve sizes from top to bottom and Mi are the weights of aggregates retained on respective sieves before and after the dispersion.

$$\text{Mean Weight Diameter (MWD)} = \sum Xi.Mi/\sum Mi \quad (8.16)$$

In eq. (8.16) and in eq. (8.17) i varies from 1 to 6 such that Xi is the mean opening size e.g. 0.05, 0.175, 0.375, 0.75, 1.5, 3.5 mm for 0-0.01, 0.1-0.25, 0.25-0.5, 0.5-1.0, 1-2, 2-5 mm size classes, respectively.

$$\text{Geometric Mean Weight Diameter (GWD)} = \exp [(\sum Mi. \ln Xi)/(\sum Mi)] \quad (8.17)$$

M13 is un-aggregated soil material less than 0.1 mm in diameter.

Measurement of Wet aggregates (Soil Conservation)

The kinetic forces of rain disintegrate soil mass. Following method is use to assess the stability of soil particles to rain energy.

1. Soil is air-dried and placed on vertically stacked sieves of 2.0 mm, 0.25 mm and a catch pan. The soil is shaken for 15 seconds on a Tyler coarse sieve shaker to separate out aggregates of 0.25 - 2.0 mm size for analysis.
2. A single layer of aggregates from 0.25 - 2.0 mm in size is spread on a 0.25 mm sieve
3. Sieve is placed at a distance of 500 mm (20 inches) below a rainfall simulator, which delivers individual drops of 4.0 mm diameter. The test is run for 5 minutes. which delivers 12.5 mm of water (approximately 0.5 inches) as drops. A total of 0.74 J of energy thus impacts the sieve over the 5 minute rainfall period.
4. The slaked soil material that falls through during the simulation, and any stones remaining on the sieve are collected, dried and weighed

The fraction of stable soil aggregates (WSA) is calculated using the following equation:

$$WSA = W_{stable} / W_{total} \quad (8.18)$$

where: $W_{stable} = W_{total} - (W_{slaked} + W_{stones})$

where: W is weight (g) with subscript stable denoting stable soil aggregates, total denoting aggregates tested and slaked aggregates are the slaked out of sieve and stones means the stones retained in sieve after test. Corrections must be made for the stones.

Measurement of Dry Aggregates

The kinetic forces of wind disintegrate soil mass, which is also a factor of soil texture, moisture content and management practices. A rotary sieve was developed to measure various size distributions of aggregates (Chepil, 1962). It simulates the wind action in the cultivated field.

Apparatus/Equipments: Nest of sieves with openings of 25, 10, 5, 2, 1, 0.5 and 0.25 mm sizes along with a pan and a lid, Rotary sieve shaker, Balance, Spade.

Procedure

1. Collect the sample when the soil is reasonably dry to avoid breakdown or change in structure with the help of a spade. Remove the plant roots or stones in the sample.
2. Keep the nest of sieves on the rotary shaker in descending order of openings. The bottom sieve is fitted in a pan.
3. Place weighed soil sample (M0) on the top sieve. Switch on the shaker for 20 minutes, and then remove the sieves, and collect the soil retained on each sieve.
4. Weigh the amount of soil retained in each sieve and note them as M1, M2, M3, M4, M5, M6 and M7.

Calculate the % distribution of dry aggregates retained on each sieve as per the following relation:

$$\text{Aggregates for a size group (\%)} > 25 \text{ mm} = (M1/M0) \times 100 \quad (8.19)$$

$$\text{Aggregates for a size group (\%) } 17.5 \text{ mm} = (M2/M0) \times 100 \quad (8.20)$$

Similar calculations for other sizes can be made with the last size being

$$\text{Aggregates for a size group (\%) } 0.375 \text{ mm} = (M7/M0) \times 100 \quad (8.21)$$

MEASUREMENT OF WATER TABLE AND HYDRAULIC HEAD

The main purpose of groundwater investigations is to get information on water logging in the area. For this purpose open wells, observation wells and piezometers are used. Normally, a network of observation wells on a grid is established. To investigate artesian pressure or hydraulic head, a cluster of piezometers at appropriate locations is installed.

Open wells

Open wells, a common feature of rural areas in India, can be used to measure the groundwater table.

Observation Wells

Observation wells are installed at strategic locations usually on a grid. Both cased and uncased wells can be used, the former being used for short-term measurements only. The uncased well is an auger hole left uncased in the field, which is employed to measure water table depth during an exploratory survey. A cased well is usually 2.5-3.75 cm diameter plastic or galvanized iron pipe. Perforations are made in these pipes throughout except for the top 0.5 m (Fig. 8.2). The perforations in the observation well are drilled in five to six rows and are covered with jute or coir or nylon rope to prevent entry of fine particles into the pipe or clogging of the tiny holes. The pipe is placed in the center of a bore hole made with a hand or power auger. The annular space between the pipe and wall of the hole over the entire length of the perforated portion is filled with gravel or coarse sand. The gravel pack facilitates the flow of water from the soil into the pipe and vice versa. The rest of the annular space is filled with fine materials (clay, clayey fine sand) whose permeability is low enough to prevent leakage along the pipe. To keep the pipe in position and to prevent any direct contact between the surface water and the backfill material, a concrete seal is provided around the pipe, extending from the ground surface to a depth of about 15 cm. The lower end is sealed while the upper end is provided with a cap, which is opened to take the observation.

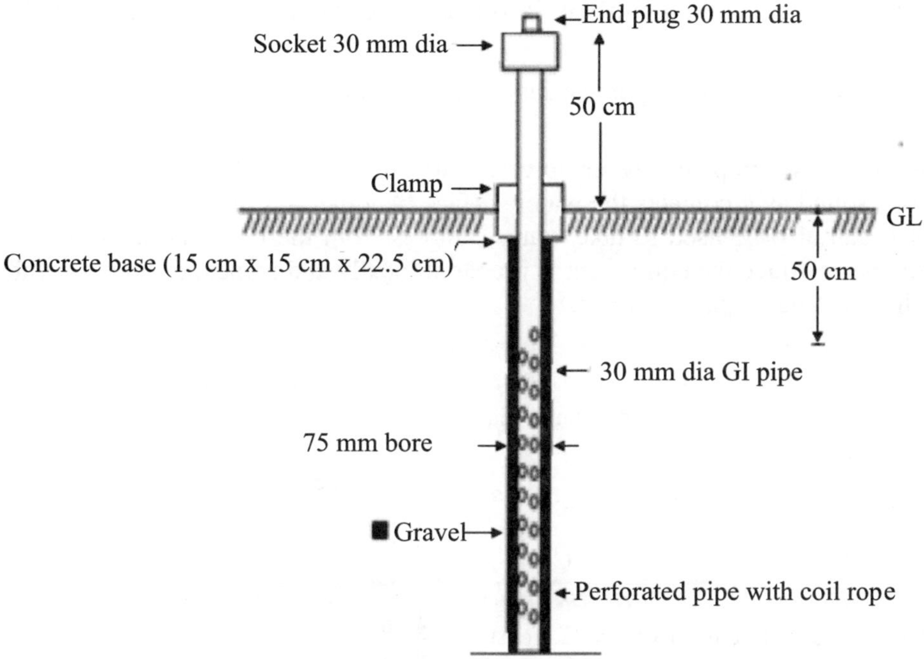

Fig. 8.2 A schematic view of an observation well (not to scale)

Piezometer

A piezometer is used to determine the hydraulic head at a particular depth. A piezometer is specialized water well. It is an open-ended non-perforated galvanized iron (or sometimes plastic) pipe of 2.5-3.75 cm diameter, driven or jetted into the ground to the depth at which the hydraulic head is to be measured. Sometimes, perforations are made only at the bottom end for about 5-15 cm length (Fig. 8.3). A battery of piezometers is normally installed at one location with the lower end of each piezometer terminating at different depths to measure the hydraulic head. The water level in the pipe corresponds with the hydraulic head at the lower end where the pipe terminates or at the center of the perforated portion of the pipe.

Measurement of Water Level

There are many ways to measure the water elevations in observation wells and piezometers. Electric water level indicators are now available for smaller diameter wells. Pressure loggers or electronic water level loggers are also used to measure the water level. Float recorders are used to record continuous variation of water levels. The simplest instrument commonly used to take observations is the chain and bell sounding device. It can be locally made as per the following method.

- Take a 1-1.5 cm diameter galvanized iron pipe piece of about 15-20 cm length.
- Seal the bottom end with a rubber cork leaving about 1 mm gap (Fig. 8.4).
- On the top, make some arrangement to attach it with a rope or a thin steel or metallic tape.

The device so prepared, when lowered into the well, makes a clearly audible "plop" sound as it contacts the water surface. Another advantage of this device is that it can also be used to take water samples from the observation well or the piezometer. Since the equipment is prepared locally and is inexpensive, a number of them can be prepared for use.

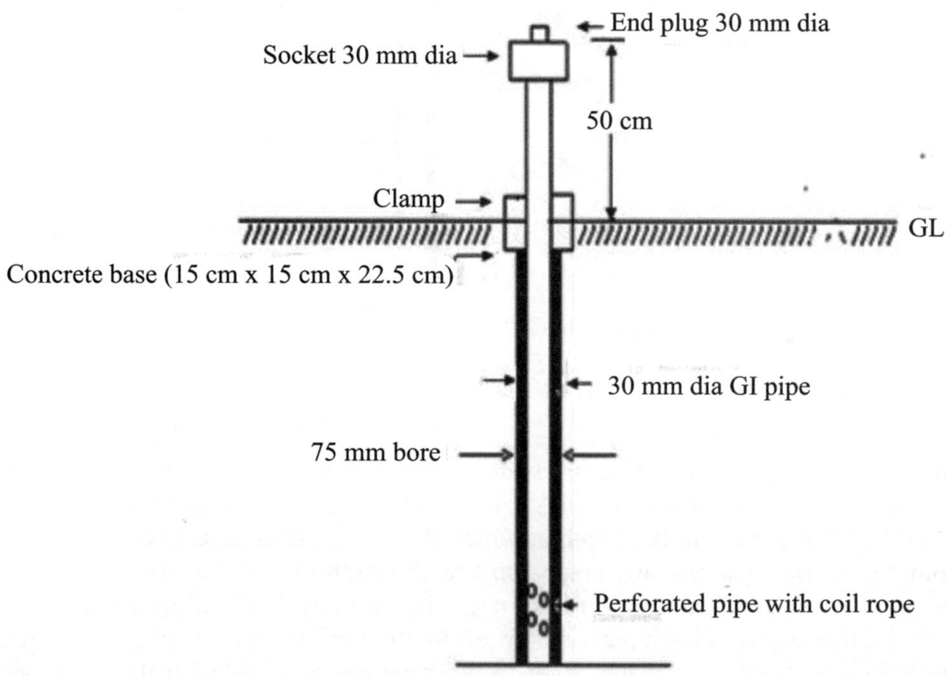

Fig. 8.3 Schematic cross-section through a piezometer (not to scale)

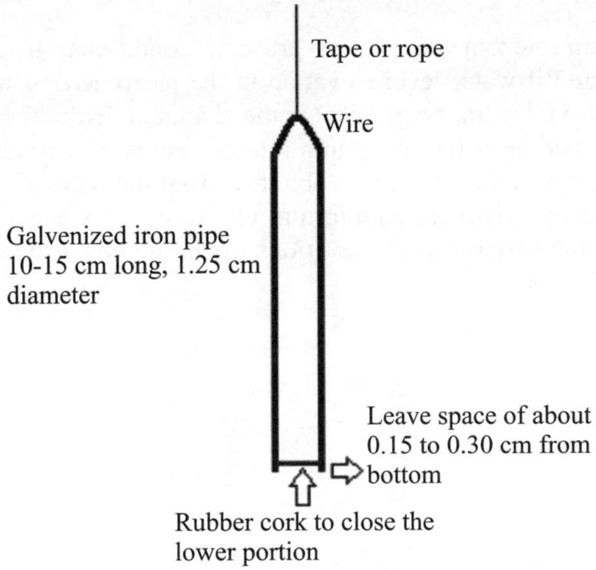

Fig. 8.4 Chain and bell sounding device to measure the groundwater levels

Analysis of Data

The data so collected is used to plot a well hydrograph. It is prepared by plotting depth to water table against time on a graph paper. A hydrographs provides insight into the causes of the rise and fall in the water level at a given site. In irrigated areas, groundwater table tends to build-up through the irrigation season, reaching a peak at the end of the season. In monsoon climatic regions, the water level is at its maximum at the end of the season (generally October) and minimum before the onset of monsoon (May or June). The hydrograph can be used to assess rise in water table during the monsoon season.

Water table contour maps show the phreatic surface (water table) and are prepared by plotting absolute water levels or elevations of the water table for all observation points on a topographical map. The map shows the relationship between ground surface and the water table. A water table contour map provides direct visual observation of the slope of the water table, horizontal groundwater movement being in the direction of the slope of the water table. Radical changes in direction or slope may indicate unusual barrier formation, perched water table, artesian conditions or high and low permeability areas leading to clues for further investigations.

Depth to water table at any point is the difference in groundwater contour and the surface contour. To know the extent and degree of water logging, an isobath map is prepared. It shows lines of equal depth to water table. This map is generally used to justify a project. It can also be used to assess the yield losses due to shallow water table when the crop yield- depth to water table relation is known.

Artesian conditions

Observations from piezometers reveal artesian conditions, if any. Artesian pressure is indicated if water level elevation in the piezometer is above the true water table (Fig. 8.5). Hydraulic gradient is the change in hydraulic head per unit distance in the direction of the maximum rate of decrease in head. The vertical component of the hydraulic gradient is obtained from the battery of piezometers by subtracting the difference in equilibrium elevations of water surface in two pipes divided by the difference in elevations of the depths at which respective piezometers terminates.

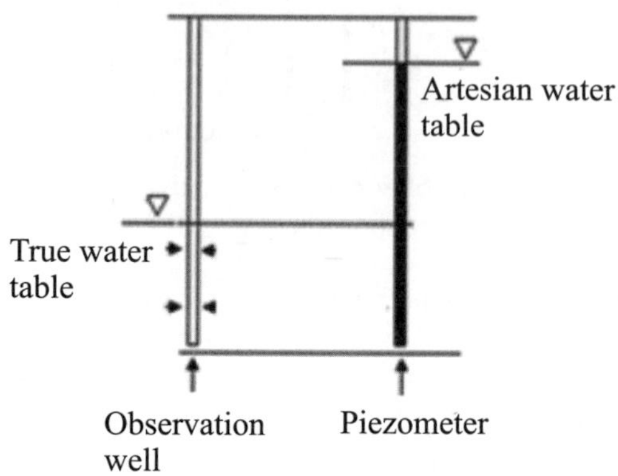

Fig. 8.5 An observation well and a piezometer showing the presence of artesian pressure

SOIL WATER CHARACTERISTICS (PRESSURE PLATE APPARATUS)

Pressure Plate apparatus provides a convenient, reliable means of measuring soil water content at different pressure/suction. The method can be used both on disturbed samples or undisturbed soil cores. Through the application of the Pressure Plate Apparatus one is able to develop moisture retention curve for any soil type. The curves relate the soil suction, at which moisture is held by the soil to its moisture content. These curves are important to know the availability of soil moisture for plant growth. For example, volumetric moisture content at field capacity is closely related to volumetric moisture content at 33 kPa suction for normal soils. For organic and sandy soils it is closely determined by the volumetric moisture content at 10 kPa suction. Permanent wilting point is closely replicated by the volumetric moisture content at 1500 kPa. The saturated cores when placed on the ceramic plates of the apparatus and subjected to an air pressure above the atmospheric pressure, the excess water is released through the microscopic pores in the ceramic plate. The smaller the pore size of the ceramic plate, higher is bubbling pressure or air entry value of the plate. A given ceramic plate can be used to extract water only below its bubbling pressure. A schematic and photographic view of the pressure plate apparatus is shown in Fig. 8.6.

Apparatus/Equipment: Pressure plate apparatus, Brass or rubber rings of 1 cm height and 6 cm diameter, Balance, Oven, Aluminium moisture cans, Pipette

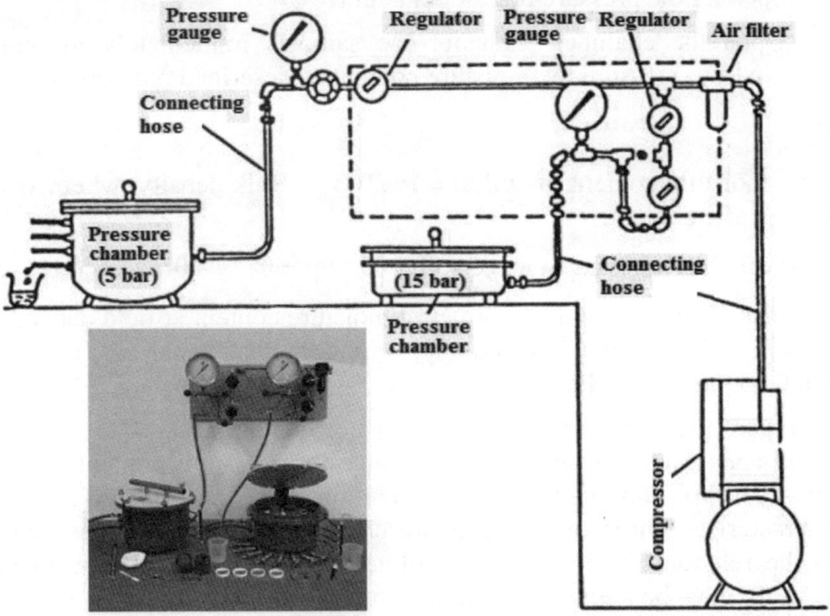

Fig. 8.6 A schematic and photographic view of a pressure plate apparatus

Procedure

1. Fill the soil samples inside the brass or rubber rings placed over appropriate ceramic plates and level them. Brass rings attached to core samplers can also be used directly for this purpose. Always test the samples in duplicate.
2. Saturate the samples by placing the plate with the rings in a trough of water for about 24 h. Keep the water level so that it is just enough to reach the upper edge of the rings.
3. Transfer the plates to the respective pressure chambers and place them on hinges in case of 5 bar extractor or on triangular support in case of 15 bar extractor.
4. Make all necessary connections and see that nylon tube and rubber sleeve is connected to the outlet pipe. Close all the unused outlets with the plug bolts and make sure that the 'O' ring is in place in the 15 bar extractor. Use a pipette or syringe to remove the excess water from the ceramic plates. Fix the lid on the pressure chambers with nuts and bolts.
5. Adjust the pressure to the required level by adjusting appropriate regulator. As the pressure builds-up inside the pressure chamber, water

comes out through the outflow tubes. Flow of water will cease as soon as soil water pressure comes in equilibrium with air pressure generally within 24 hr. At this time release the pressure in the chamber by opening the pressure release regulator.

6. Open the chamber. Transfer the samples immediately to moisture boxes to measure the moisture content as described previously.

Analysis

Volumetric moisture content of soil, $\theta = (w/100) \times$ Bulk density, where w is the gravimetric water content.

$$\text{Available water capacity of soil} = \theta f - \theta p = \theta_{0.033} - \theta_{1.5} \quad (8.22)$$

Such that θf and θp are the volumetric moisture content at field capacity and at permanent wilting point respectively, which is also equal to the moisture content at 0.033 and 1.5 MPa suction.

To prepare the water retention curve, water content at 0, 10, 30, 100, 345, 1033, 3100 and 15000 cm are obtained. Once the moisture content at various pressures is known, draw the soil water retention curve, also known as the soil-water characteristic curve or the moisture characteristic curve. Water retention curve is the relationship between the volumetric water content, θ, and matric suction, ψ. It may be seen that soil moisture characteristic curve is strongly affected by soil texture (Fig. 8.7). Greater the clay content, the greater the water content at any particular suction and more gradual the slope of the curve.

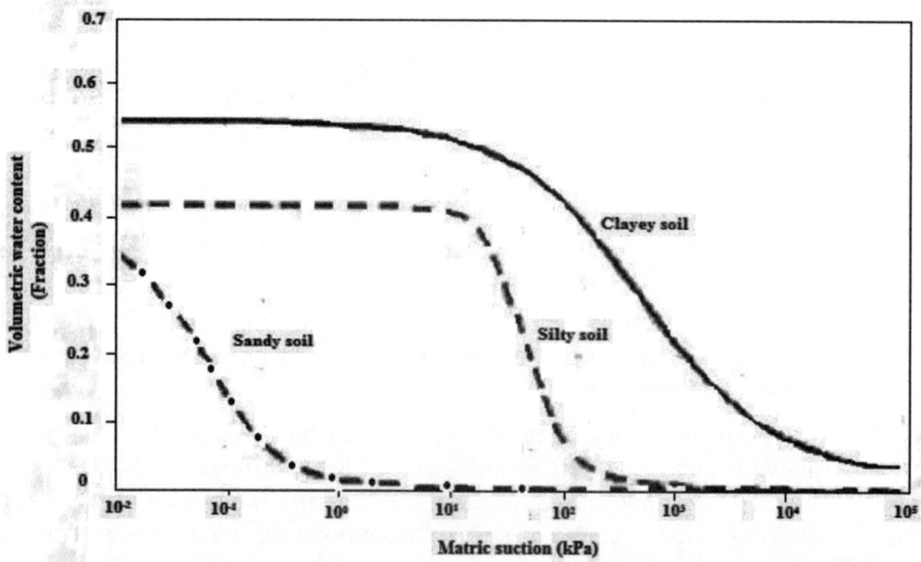

Fig. 8.7 A schematic view of the water retention curves of 3 different soil textures

Precautions: Increase pressure in the pressure chamber gradually until the desired level is reached. Take all precautions to ensure that there is no air leakage from the pressure plate assembly. Before opening the lid, assure that the pressure is fully released.

INFILTRATION RATE

The infiltration rate, a property of surface layer, is the rate at which water enters into the soil. It is usually measured by the depth (mm) of the water that enters the soil in a unit time usually hour or a day. It is primarily the vertical rate of water movement into the soil surface (one-dimensional). The cylinder or ring infiltrometer is commonly used to measure infiltration rate.

Apparatus: Double ring infiltrometer, Point or hook gauge or Marriott arrangement, Source of water, Timer, Plastic or polythene sheet

Cylinder Infiltrometer

A cylinder infiltrometer, 20 to 30 cm in diameter and 30 to 40 cm in length, is used to measure the infiltration rate. It is made of smooth steel (2 mm thick), rigid enough to be driven into the ground, but thin enough to enter the soil with minimum disturbance. It may be reinforced around the upper edge. A buffer cylinder (together called a double ring infiltrometer) having a diameter at least 30 cm greater than the inner cylinder and of about the same or even shorter length is used. Alternately, a buffer ring of earth can be made surrounding the inner cylinder. It is necessary to make the flow in the inner cylinder vertical (Fig. 8.8). The desired vertical flow from the inner cylinder is only possible with a double ring infiltrometer.

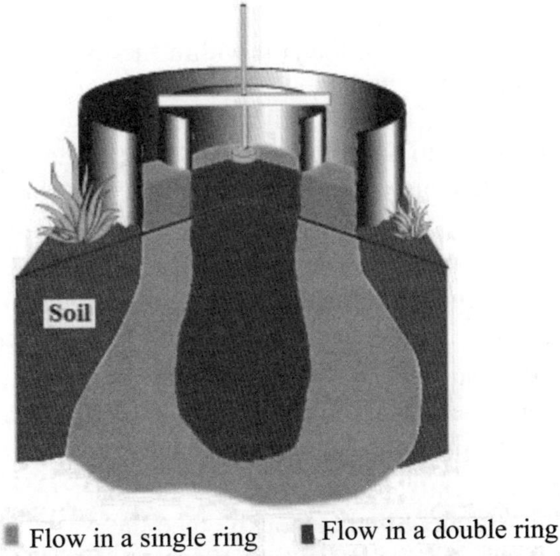

Fig. 8.8 Anticipated schematic flow in a double ring infiltrometer (inside bulb). Outside bulb is anticipated flow from single ring infiltrometer

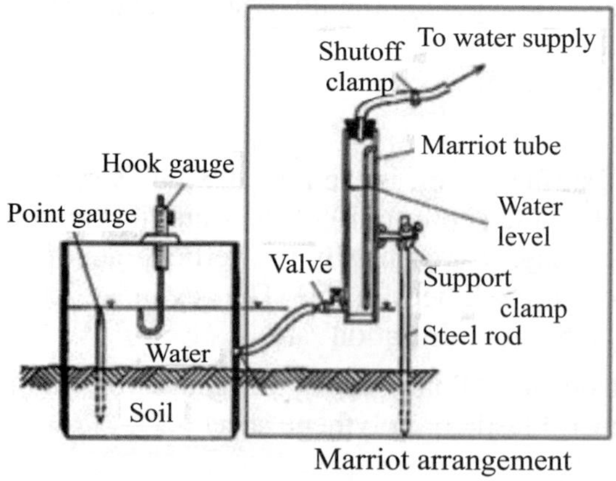

Fig. 8.9 Measurement of infiltration rate schematic view (left) and a photographic view (right)

Procedure

1. Select a representative location for each cylinder and examine carefully for signs of unusual surface disturbance, such as stones, cracks which might give non-representative readings, etc.
2. The two metal concentric rings (or cylinders) are gradually hammered into the soil causing least disturbance to the soil surface. The cylinder should be installed as vertically as possible (Fig. 8.9 right). Drive the infiltrometer cylinder to a depth of at least 15 cm. The outside cylinder need not be driven as deep as the inner cylinder, may be limited to 10 cm.
3. Place a plastic sheet on the soil within the inner cylinder. The plastic should be in contact with the soil at the bottom of the infiltrometer and extend up the walls of the infiltrometer at least 15 cm. Fill the buffer pond with water to a depth roughly equal to the depth desired in the inner ring. Maintain roughly equivalent depths throughout the period of observation.
4. Fill the cylinder infiltrometer with water to a depth of about 10 cm. In extremely porous (high intake) soils, a greater depth may be needed.
5. Place the hook gauge board on the cylinder and set the point of the gauge at the water level. Other arrangement such as a point gauge or a Marriott arrangement to ensure a constant head can also be made (Fig. 8.9 left). Read the scale from the top edge of the clip.
6. Quickly, but gently, remove the plastic sheet recording the time at which this is done. Simultaneously, take the water level reading in the inner cylinder. It is the initial time reading.
7. Take readings of the hook gauge at periodic intervals and record the time and water level readings. Intervals between observations should

be short (2 to 5 min) at the beginning of the test. After 20 minutes the intervals may be increased to 10 to 20 min. After about 2 hours, measurements may be made at 30 to 50 minute intervals.

8. When the water level has dropped 4 to 5 cm in the cylinder infiltrometer, carefully add a sufficient volume of water to return the water surface to the approximate initial level. It may not be needed if constant head arrangement is made. The known volume of water added, divided by the cross-sectional area of the cylinder, gives the depth of water added to the cylinder.
9. Where abnormally high or low infiltration values are indicated by the test results, the infiltrometer should be dug out and the soil examined for possible causes. If an obvious cause for a high or low rate of infiltration is seen, the test should be repeated at another nearby place.
10. The measurements should be continued until the change in depth over time becomes fairly constant. The time required will generally vary from 3 to 6 hours, depending on the pre-existing moisture conditions and soil hydraulic conductivity.

Observation and Analysis Sheet

Data are recorded as per the given sheet (Table 8.2). Clock time is recorded in column 2 and is used to calculate the time interval and cumulative time (Column 3 and 4). Depth of water is filled in column 4 to find out the intake of water during the time interval. Column 6 and 7 are than filled using calculations as described in these columns.

Table 8.2 Note sheet for collection and partial analysis of data

Sr No.	Clock time	Time interval (min)	Cumulative time (min)	Depth of water (mm)	Intake (mm)	Cumulative infiltration (mm)	Infiltration rate (mm h^{-1})
	1	2	3	4	5	6	7
1	T1	0	0	A1	0	0	0
2	T2	Row (2-1) from column 1	Row (1 +2) From column 2	A2	Row (2-1) from column 4	Row (1+2) from column 5	(Column 5/Column 2)x 60
3	T3	Row (3-2)	Row (1+2+3)	A3	Row (3-2)	Row (1+2+3)	
4							
5							
6							
n	Tn		Row (1+2+3…n)	An	Row (n - (n-1)) from column 4	Row (1+2+3….n) from column 5	(Column 5/Column 2)x 60

On a normal arithmetic paper, plot the data of column 3 on X-axis and column 7 on Y-axis. It will give infiltration rate curve. The constant rate towards the end of the curve is the basic infiltration rate (mm h^{-1}).

On a normal arithmetic paper, plot the data of column 3 on X-axis and column 6 on Y-axis. It will give cumulative infiltration curve. It can be used to measure the total water infiltrated in a given time.

HYDRAULIC CONDUCTIVITY

Hydraulic conductivity is defined as the rate of flow of water through a porous medium under laminar conditions through a unit cross-sectional area perpendicular to the direction of flow under unit hydraulic gradient and under standard temperature conditions. A number of methods to measure hydraulic conductivity are in vogue. Few of them are being included in this section.

Constant Head Method

The constant-head test is one of the most commonly used methods to determine the saturated hydraulic conductivity of coarse-grained soils. The test is based on the direct application of Darcy's law to a soil liquid configuration. It represents a one-dimensional steady flow of a percolating liquid through a saturated column of soil of uniform cross-sectional area. In this method, a constant head is maintained at the soil surface of a saturated sample and water is allowed to flow through the sample till the measured outflow attains a constant value. Different kinds of instruments are available but local arrangements can also be made to prepare a set-up in the laboratory (Fig. 8.10 left).

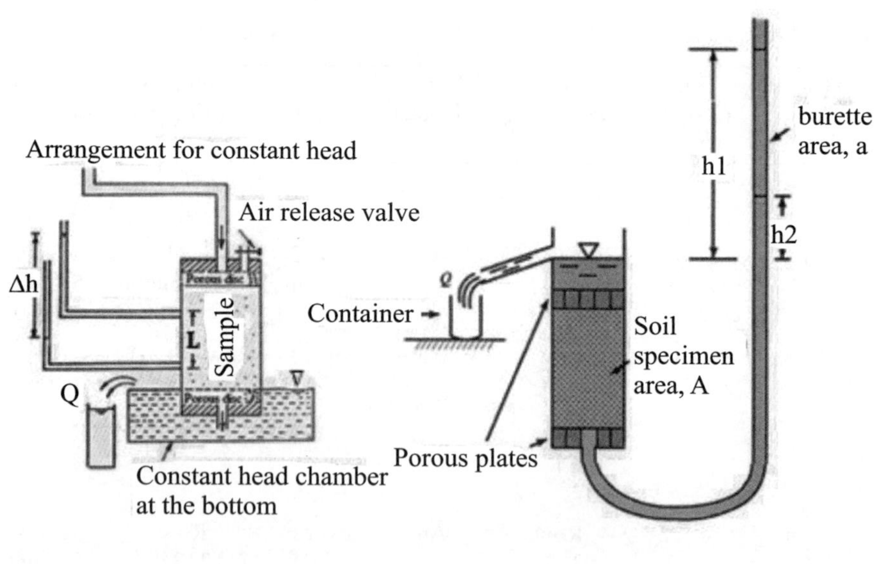

Fig. 8.10 Schematic diagram showing the basic set-up of constant head (left) and variable head permeameter (right)

Apparatus/Equipment: Constant/variable head permeameter(s), Core sampling kit for undisturbed soil sampling, permeability stand with funnels, Perforated brass plates, Water reservoir with Marriott arrangement, stop watch, Graduated cylinder, Knife, Muslin cloth, Filter paper, Rubber rings or tape

Procedure

Disturbed soil sample

1. Measure the internal dimensions of the mould. Weigh the mould with dummy plate to the nearest gram. Apply a little grease on the inside of the mould. Clamp the mould between the base plate and the extension collar and place the assembly on a solid base.
2. Take about 2.5 kg of the soil sample, from a thoroughly mixed soil. Compact the soil at the required dry density using a suitable compacting device. If the soil profile is layered, ensure to take separate samples for each layer and fill in the same fashion. The idea is to create real field situation of the soil profile.
3. Remove the collar and base plate. Trim the excess soil level with the top of the mould. Clean the outside of the mould and the dummy plate. Find the mass of the soil in the mould and calculate the bulk density.
4. The mould with the sample is placed on a porous plate covered with filter. The plate is now placed over the permeameter. Connect the specimen to the top inlet of the constant head reservoir.
5. The bottom outlet is opened and water is allowed to flow until a steady flow is established. Measure the steady state flow for a given time interval say t minutes.
6. Measure the difference of head (h) in levels between the constant head reservoir and the outlet at the base. Sometimes, two manometers are provided at a known distance separated at a distance, L. The readings of these manometers will also give the value of Δh. If no manometers are provided, then L is the length of the soil column and Δh is the head difference from the outlet to the constant head level. If a constant head level is maintained at the soil surface (In most local set-ups, it is the case), then use $L/(L+h)$ instead of $L/\Delta h$, such that h is the water head above the soil surface.
7. Take about 3 reading for the same interval and average out the values. The 3 values should be nearly same, being steady state. If there are large differences, allow more time and take flow readings after some more time.

Undisturbed soil samples

Clear the soil surface of leaves and trashes. Push the core (permeameter) gently into the soil and obtain an undisturbed soil core. Trim the excess soil on both ends. Put lids on both ends so as not to disturb the soil during transport to the laboratory. Rest of the procedure is same as in the case of undisturbed soils

Observations and Calculations

By measuring the volume Q of the water that flows through the system during time *t*, the saturated hydraulic conductivity K of the soil is determined as follows:

$$K = QL/(\Delta h \times t \times A) \tag{8.23}$$

Here, K is coefficient of permeability, Q is volume of water discharged, L is difference between water level in the two manometers, A is cross-sectional area of specimen, t is total time in which volume of water is collected and Δh is difference in head. Use similar units for the parameters in eq. (8.23).

Falling Head Method

The falling head test with the permeameter is primarily used to determine K value of fine-grained soils. Like the constant-head method, the falling-head test also operates in accordance with direct application of Darcy's law to a one-dimensional, saturated column of soil with a uniform cross-sectional area. The falling-head method differs from the constant-head method in that the liquid that percolates through the saturated column is kept at an unsteady-state flow regimen in which both the head (Fig. 8.10 right) and the discharge vary during the test. Prepare the soil specimen as before and use the variable head apparatus to measure the hydraulic conductivity. Follow a similar procedure as outlined for constant head permeameter test. By measuring the change in head in the standpipe from h_1 to h_2 during a specified interval of time *t*, the saturated hydraulic conductivity is determined as follows (Klute and Dirksen, 1986):

$$K = [aL/A \times \Delta t]\,[\ln (h_1/h_2)] \tag{8.24}$$

Here a is cross sectional area of the stand pipe, L is length of soil column, A is cross section area of soil column, t is time interval for the head to drop from h_1 to h_2 at the end of test.

Interpret the results as per the information provided in Table 8.3.

Table 8.3 Interpretation of results

Hydraulic conductivity (cm h^{-1})	Classes
<0.125	Very slow
0.125-0.5	Slow
0.5-2.0	Moderate slow
2.0-6.25	Moderate
6.25-12.5	Moderately rapid
12.5-25.0	Rapid
>25.0	Very rapid

Auger Hole Method

The auger-hole method is a simple, rapid, and reliable method to measure hydraulic conductivity (HC) in the field. The method is used to measure HC of the soil below a water table. The method, originally proposed by Diserens (1934), has been improved over time by various workers, the most important modification coming from Hooghoudt (1936). The general principle is to bore an auger hole to a certain depth below the water table. When equilibrium is reached with the surrounding groundwater, a part of the water in the hole is removed so that water seeps into the hole again. The rate at which the water rises in the hole is measured and converted to the hydraulic conductivity using an appropriate formula. The method is most commonly used in the design of drainage systems in waterlogged land and in canal seepage investigations. For more details of this and inverse auger hole methods, readers may refer to Gupta (2019).

Equipment: Soil auger of suitable diameter normally more than 6 cm but less than 14 cm, Extension rods, Bailer or pump, Float gauge set consisting of measuring tape and a float

Procedure

1. Make the auger hole, note down the depth where the groundwater level is reached for the first time. Drill the hole going about 60-90 cm below the groundwater table (Fig. 8.11). It may be useful to observe the changes in soil characteristics (color, water saturation, etc.).
2. Allow equilibrium with the surrounding groundwater. Measure the stable groundwater level.
3. Remove the water from the hole 2 to 3 times to open the soil pores. Finally, remove the water as quickly as possible so that the water level is 20-40 cm below the static water level.
4. Measure the rate of the rise of water level at frequent preferably at regular intervals. One can use the float gauge with a measuring tape or any electrical device to measure the water levels. Take about 8-10 readings. In this process see that not more than 25-30% of the water removed is recouped at the time of last measurement.

Equations

Equations given by Ernst (1950) are now commonly used to determine the hydraulic conductivity using the data from an auger hole test.

For D > H/2, hydraulic conductivity is calculated by the relation (van Schilfgaarde, 1974);

$$K = \frac{4000\, r^2}{(H + 20\, r)[2 - (y/H)]\, y} \times \frac{\Delta y}{\Delta t} \tag{8.25}$$

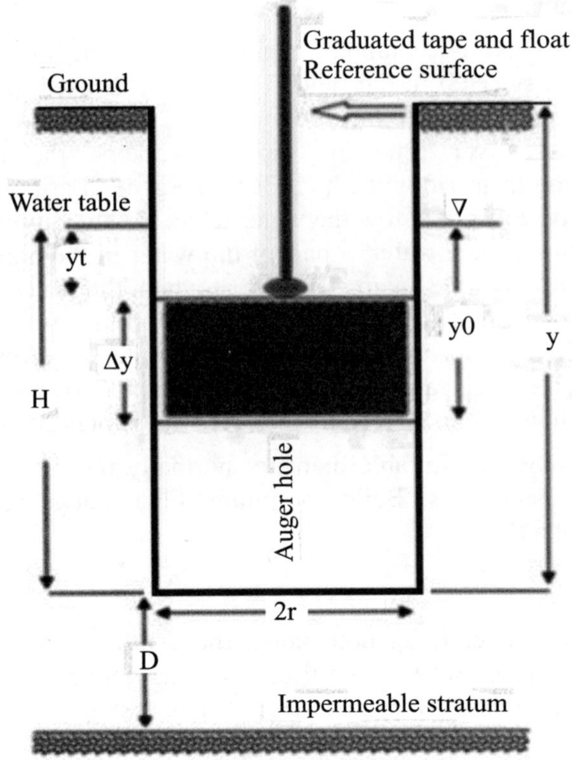

Fig. 8.11 A schematic view of the auger hole to measure hydraulic conductivity

Here K is in m/d, r, H, y and Δy in cm and Δt in seconds. The parameter y is average depth of the water table in the auger hole given by $y_0 - ½\ \Delta y$.

When D = 0, the auger hole rests on the impermeable layer and equation used to estimate hydraulic conductivity is:

$$K = \frac{3600\, r^2}{(H + 10\, r)[2 - (y/H)]\, y} \times \frac{\Delta y}{\Delta t} \quad (8.26)$$

The parameters are the same as described for Eq. (8.25). For intermediate values, still there is a lack of a suitable formula. In such events values from these two equations are worked out to select an intermediate value or Eq. (8.26) is used.

To obtain acceptable accuracy the following conditions should be met with.

2r > 6.25 cm and < 14 cm

H > 20 cm and < 200 cm

y > 0.2 H

$\Delta y < 1/4\ y0$

$S > H$

Example

Borehole characteristics

Radius of the auger hole (cm), r = 4 cm (More than 6.25 cm)

Depth to water table from reference of ground surface = 49 cm

Depth of the auger hole below water table, H = 101 cm

Depth of impermeable layer below auger hole, D = 70

Time t =50 s

$yt = 42.8,\ y0 = 56.5,\ \Delta y = 56.5 - 42.8 = 13.7$

$y = 56.5 - 13.7/2 = 49.65$ cm

$r/y = 0.08$

$H/r = 25.5$

$y/H = 0.49$

Field observations are reported in Table 8.4 and calculations are performed as follow.

Since D > H/2, we use eq. (8.25) to calculate K

Table 8.4 Water level observations from reference, y't and below the water table, yt

t (sec)	y't	Reference point (cm)	yt	yt- yt-1
0	105.50	49	56.5	- -
10	102.50	49	53.5	3.00
20	99.10	49	50.1	3.40
30	96.20	49	47.2	2.90
40	93.30	49	44.3	2.90
50	91.80	49	42.8	1.50
60	89.70	49	40.7	2.10
70	87.50	49	38.5	2.20
80	85.60	49	36.6	1.90
90	83.90	49	34.9	1.70
100	82.30	49	33.3	1.60

$$K = \frac{4000\, r^2}{(H + 20\, r)\,[2 - (y/H)]\, y} \times \frac{\Delta y}{\Delta t}$$

$$K = [(4000\times4^2)/(101+20\times4)\ (2- 0.49)\ \times 49.65] \times (13.7/50)$$

$$K = 1.29 \text{ m d}^{-1}$$

Inverse Auger Hole Method

Another variation of the auger hole test is the inverse auger hole method. It is performed in areas where water table is deep at the time of test. There are two variation of this test.

- Constant head method
- Variable head method

The later test is more commonly applied in field investigations because it dispenses with the need to keep a constant water level in the hole. Therefore only this variation is included in the Handbook. The basic principle and few parameters that are to be measured are shown in Fig. 8.12.

Equipment: Soil auger of suitable diameter normally more than 6 cm but less than 14 cm, Extension rods, Bailer or pump, Float gauge set consisting of measuring tape and a float.

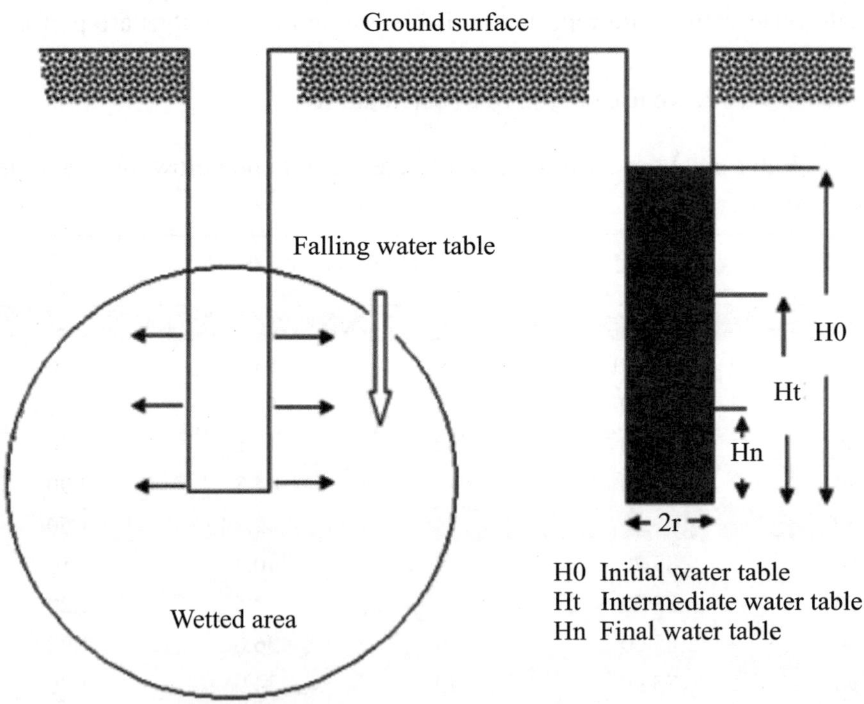

Fig. 8.12 Inverse auger hole method to determine hydraulic conductivity, wetted area (left) and parameters of measurement (right)

Procedure

1. Make the auger hole to the desired depth of measurement. It is useful to observe the changes in soil characteristics (color, water saturation, etc.).
2. Fill the water in the auger hole and let it be absorbed. Repeat the process few times so that sufficient area around the auger gets saturated and infiltration attains more or less a constant value (Fig. 8.12 left).
3. Fill the water in the auger and take readings of the falling water table in the auger at regular interval as in the case of auger hole method described in the previous section. A float gauge with a measuring tape or any electrical device can be used to measure the water level in the auger. Take about 8-10 readings. The Performa to collect data is shown as Table 8.5. The hydraulic conductivity is then calculated by the equation given in Ritzema (1994).

Table 8.5 The Performa to arrive at the values of (h_t + r/2)

Time (s)	Htt (cm)	H_t = D- Htt	H_t + r/2
	Water level in the bore from top of the auger hole or land surface	D is the depth of the auger hole from top or ground surface, subtract values in column 2	Add r/2 in values of column 3
0		H_0	H_0+ r/2
t1			
t2			
tn			

$$K = 1.15\,r\,\frac{\log(H_o + r/2) - \log(H_t + r/2)}{(t - t_o)} \quad (8.27)$$

$$= 1.15\,r\tan\alpha \quad (8.28)$$

Here H_o is the initial height of the water column in the hole (cm) at time t_o, H_t is the water level (cm) in the auger at any time t, r is the radius of auger hole (cm), t and t_o are the times (s) since the beginning of the test and K is the hydraulic conductivity (cm s^{-1}). The most appropriate way to use eq. (8.28) is to plot (H_t + 0.5 r) against t on a semi logarithmic paper, the former being on the log scale. It will result in a straight line as shown in Fig. (8.13). The slope of this line is equal to tan α, which is used in eq. (8.28) to obtain the value of K in cm s^{-1}. Multiplication of this K value by 864 gives K in m d^{-1}.

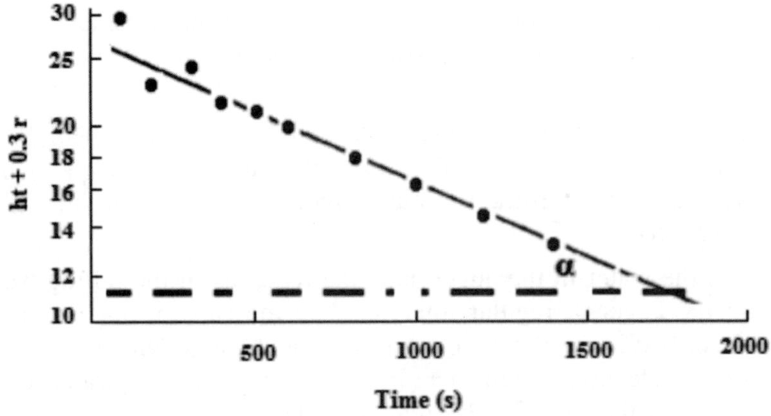

Fig. 8.13 Plot of data against time for inverse auger hole method

DISPERSION

Vulnerability of the soils to get eroded is referred to as erodibility of soils. Although a function of both the physical characteristics of soils and the land management practices, soil texture (sand, slit, and clay composition), organic matter content, structure and permeability in addition to iron, aluminium and sodium contents are a few of the major factors that affect soil erodibility. Dispersion, often seen as 'muddy' or 'milky' water in surface water resources or runoff, indicates the potential of soil to erode. It is commonly encountered in alkali soils due to their high exchangeable sodium due to swelling of clay platelets resulting in collapse of clay aggregates.

Crumb Test

Equipment: Glass jar or dish, Distilled water

Field testing for dispersive soils can be conducted by observing the behaviour of air dried aggregates in distilled water or rainwater.

1. Collect soil aggregates (1-2 cm diameter) from any soil layer in the soil profile. If moist, dry the aggregates for a few hours until air-dried.
2. Place the aggregates in a shallow glass jar or dish containing distilled water or rain water. Leave the aggregates in water without shaking or disturbing them for 2 hours.
3. Observe and record the milky ring around the aggregates. Don't bother if the soil collapses or bubbles are formed. Qualitative conclusions are drawn as per guidelines in Table 8.6.

Note: Crumb test gives a good indication of the potential erodibility of clay soils. However, a dispersive soil may sometimes give a non-dispersive reaction in the crumb test.

Table 8.6 Guidelines on categorization of soil type

Observation	Soil type
Water remains clear although particles may crumble. Boundary of crumbs clearly defined	Non-dispersive soil
Discolouration of surrounding particles or distinct cloudiness surrounding some. Boundary of crumbs vaguely defined.	Slightly dispersive
Discolouration and cloudiness surrounding most or all particles. Boundary of crumbs not able to be defined.	Dispersive
Discolouration and cloudiness extending vertically throughout the water	Highly dispersive

Laboratory Method

Dispersion coefficient defined as the ratio of water dispersible clay to the total clay expressed as percent is usually high in alkali soils. Compared to normal soils where dispersion coefficients may be in the range of 20-30%, it may exceed 90% in alkali soils having pH more than 10.0. In soil conservation, dispersion coefficient is defined as the ratio of the soil material <0.005 mm after limited mechanical dispersion without dispersants to the total material <0.005 mm expressed as a percentage. The test to determine dispersion coefficient is conducted in a similar manner as the particle size analysis. Therefore, the test can be conducted along with the test on particle size analysis (Kinney, 1979, Craze and Hamilton, 1991, Kondel, 1991).

Apparatus and Reagents: Same as in particle size analysis using hydrometer.

Procedure

1. Measure the clay content of the soil as in the case particle size analysis
2. Measure the clay content with the same method except without adding the dispersant (only adding the deionized water) and shaking only for about 10 minutes. Take hydrometer readings at the same time as in item 1.
3. Fill a 1000 mL cylinder with deionized water. Fill another 1000 ml cylinder with deionized water and 20 mL of dispersant (25% sodium hexametaphosphate). Take hydrometer readings at the same time as in item 1.

$$\text{Dispersion (\%)} = 100\,(Rl - Rw)/(Rp - Rd) \tag{8.29}$$

Here Rl is the hydrometer reading of soil suspension without addition of dispersant and with limited mechanical dispersion, Rp is the hydrometer reading of the soil suspension with dispersant and prolonged mechanical dispersion, Rw is the hydrometer reading of deionized water, and Rd is the hydrometer reading of deionized water plus dispersant . Use the following guidelines to characterize the soil.

Percent dispersion	Degree of dispersion
< 30	Non-dispersive
30-50	Intermediate
>50	Dispersive

■■■■

GLOSSARY OF TERMS

Acid: A substance that has a pH value between 0 and 7; a substance that has more free hydrogen ions, H^+, than hydroxyl ions, OH^-.

Adsorption: It is the adhesion of atoms, ions or molecules from a gas, liquid or dissolved solid to a surface. This process creates a film of the adsorbate on the surface of the adsorbent e.g. exchangeable cations and anions on soil particles. It is different than absorption, which involves the whole volume of the material.

Adjusted R_{Na}: A parameter similar to SAR describing a alkali water to take care of precipitation of Ca due to presence of HCO_3. In this case, Ca concentration of the water is adjusted for the HCO_3 content and the EC of the water.

Adjusted SAR: A water quality parameter that is nearly similar to SAR.

$$\text{Adj. SAR} = \text{SAR}\ (1 + 8.4 - pH_c)$$

The values of pH_c above 8.4 indicate a tendency to dissolve lime from the soil and values below 8.4 indicate a tendency to precipitate lime from the applied water. The term is now obsolete.

Aggregate: A unit of soil structure usually formed by natural processes and generally < 20 mm in diameter. Aggregation is the process in which primary soil particles (predominantly silt and clay) combine to form secondary units (aggregates or clods).

Alkali soil: A soil that contains sufficient exchangeable sodium (ESP > 15) and $pH_s > 8.2$ to interfere with the growth of most crop plants, without appreciable quantities of soluble salts (EC_e is less than 4.0 dS m^{-1} but now the limit is 2.0 dS m^{-1}). This soil is also called sodic soils.

Alkaline: Water or soils containing an amount of alkali (strongly basic) substances sufficient to raise the pH value above 7.0.

Alkaline soil: A soil with a pH above 7.

Alkali water: Water having variable SAR but RSC greater than 2.5 meq L^{-1}. EC may or may not be high. If SAR is more than 15, it is called high SAR alkali water. Also called sodic water.

Alkalization: A process whereby the exchangeable-sodium content of the soil is increased. The process is also called sodification or sodication.

Amendment (soil): An alteration in the properties of a soil by the addition of substances such as lime, gypsum and organic materials to make the soil more suitable for the growth of plants; Any substance used for this purpose.

Anion: Anions are negatively charged particles (ions) such as chloride (Cl^-), sulphate (SO_4^{2-}), carbonate (CO_3^{2-}), or bicarbonate (HCO_3^-).

Anion exchange capacity: The total amount of exchangeable anions that a soil can absorb. It is expressed as milli-equivalents per 100 g of soil.

Available water: The portion of water in a soil that can be readily absorbed by plant roots. Defined as the quantity of water retained in the soil between its field capacity (FC) and permanent wilting point (PWP). Usually irrigation is applied when plants have used about half of the available water.

Bar: A unit of pressure. One bar equals 14.5 pounds per square inch (psi) or about 0.987 standard atmospheres or 10^5 Pa or 100 kPa.

Base: A substance that in aqueous solution release hydroxide (OH^-) ions, have a pH greater than 7.

Base Exchange capacity: See cation exchange capacity.

Basic infiltration rate: The nearly constant infiltration rate of the soil achieved after a reasonable time depending upon the soil texture.

Black alkali: An alkali soil often referred as such because of black patches that appear due to accumulation of organic matter at the soil surface. The term is now almost obsolete.

Blending: Also referred as mixing; a term used in conjunctive use of multi-quality waters when two or more qualities of waters are mixed for irrigation.

Bulk density: It is defined as the dry weight of soil per unit volume of soil. Bulk density considers both the solids and the pore space. It is expressed in g cm^{-3} or kg m^{-3}.

Calcareous soil: Soil containing sufficient free calcium carbonate or calcium-magnesium carbonate to effervesce visibly when treated with cold 0.1 *N* hydrochloric acid.

Calcite: Calcium carbonate ($CaCO_3$). A trade name for finely ground limestone, very high in calcium carbonate, which is used to neutralize soil or water acidity.

Calcium: An important cation in soils and an important plant nutrient. Calcium is divalent cation and attracted strongly to clay and/or organic matter. This attraction means that calcium holds clays together. It causes flocculation (clumping together) of wet clays resulting in good soil structure. Gypsum and lime are both calcium salts often used to obtain this effect.

Capillary fringe: The zone above the free water table where some or all of the capillary interstices are filled with water; this water is continuous with the water in the saturated zone, but is held above that zone by capillarity against gravity.

Capillary rise: The upward movement of water from a free water table due to adhesion of water to the tubular soil pores (capillaries) and the cohesion of water molecules.

Cation: Cations are positively charged particles of atoms or molecules. (Ca^{2+}), magnesium (Mg^{2+}), sodium (Na^+), potassium (K^+), or ammonium (NH^{4+}). The amount and the ratios of various cations play an important role in plant nutrition and in maintaining the soil structure.

Cation exchange: The interchange between a cation in solution and another cation on the surface of any surface-active material such as clay colloid or organic colloid.

Cation exchange capacity (CEC): A measure of the net negative charge of a soil. Measured as the total quantity of cations that can be exchanged on a unit of soil material, expressed as milliequivalents per 100 grams of soil (meq 100 g^{-1}), millimoles of charge per kilogram of soil (mmol [+ charge] kg^{-1}) or centimoles of charge per kilogram of soil (cmol [+ charge] kg^{-1}).

Concentration: Quantitative amount of a solute, chemical or pollutant in a specified volume or weight of solvent, air, water, soil or other medium. Accumulating a level of some material over and above the level found in the ambient environment, generally applied to an organism.

Conjunctive use: Use of multi-quality, multi-sources water in conjunction to get better returns in comparison to when these waters are used separately.

Crop tolerance: Tolerance of crops to adverse abiotic stresses such as water logging, soil salinity, sodicity, heat and cold waves etc.; it may be to the biotic stresses as well.

Crust: A surface layer of soil, from a few millimeters to 2.5 cm thick, caused by rain drops impact, clay dispersion and clay translocation. When dry, it is much more compact, hard, and brittle than the material just under it.

Cyclic Use: A term used in conjunctive use of multi quality waters when two or more qualities of waters are used to irrigate the crop in some cycle or sequence to avoid stress at salt sensitive stages.

Degradation: The process of soil change from one type to another resulting in increased production constraints. Processes of soils being salinized or sodified or even the replacement of sodium by hydrogen during leaching.

Desalinization: Removal of excessive soluble salts from the saline soil, usually by leaching.

Dispersed soil: Soil in which clay readily forms a colloidal solution, usually have low permeability, tend to shrink, crack and become hard on drying and become plastic on wetting.

Dispersion: Dispersion is the breakdown of soil clay minerals into single particles in solution. The opposite is flocculation. Dispersion occurs when the bonds in a soil are not strong enough to keep the particles together. It is usually because a soil has too great a proportion of exchangeable sodium in it, or is too low in organic matter. However, working a soil too wet or too dry can also increase its tendency to disperse on wetting.

Dispersion coefficient: It is defined as the ratio of water dispersible clay to the total clay expressed as percent.

Dissolved solids: Amount of minerals, such as salt, that are dissolved in water; amount of dissolved solids is an indicator of salinity or hardness. Expressed as total dissolved salts (TDS)

Distillery spent wash: A byproduct of the molasses based distilleries, nutrient rich wastewater with high biological oxygen and chemical oxygen demand (BOD and COD). Used as an amendment to reclaim alkali soils and water.

Drainage: Unimpeded downward movement of water beyond the root zone. If impeded, it should be artificially strengthened by human actions.

Drainage system: Collection of open and/or closed drains, used to collect and dispose of excess surface or subsurface water.

Edaphic: Of or pertaining to the soils; is the condition resulting from or influenced by factors inherent in the soil or other substrata, rather than by climatic factors.

Electrical conductivity: It is the property of a medium expressing its capacity to carry electrical charge from one point to another point. It is the reciprocal of the electrical resistivity. The resistivity is the resistance in ohms of a conductor which is 1cm long and has a cross-sectional area of $1cm^2$. This property, usually expressed in milli or micro ohm per cm, is used to express the salinity of the soil extracts, drainage water, irrigation water, etc. International units of EC are deci-Simens per meter or dS m^{-1}. 1 dS m^{-1} is equal to 1 mmhos cm^{-1}.

Elemental sulphur (S): A yellow, inert crystalline mineral that is finely ground. In soil, elemental S is oxidized to sulphuric acid via microbial activity. The rate of elemental S oxidation in soil is most rapid in warm, moist soils. Elemental sulphur as an amendment is applied whenever sodium is in excess and there is an excess of lime.

Equivalent weight: The mass of a substance especially in grams that combines with or is chemically equivalent to eight grams of oxygen or one gram of hydrogen **:** the atomic or molecular weight divided by the valency.

Excess lime (Calcium carbonate): Free lime; an excess of lime (1.0% or more) reduces the availability of phosphorus and micronutrients and will cause the soil to crust. Nearly all soils with an excess of lime will have pH values between 7.0 and 8.4.

Exchangeable-Cation Percentage (ECP): The extent to which the adsorption complex of a soil is occupied by a particular cation. See Exchangeable Sodium Percentage.

Exchange complex: Active surface constituents of soils both inorganic and organic which are capable of ion exchange.

Exchangeable Sodium Percentage (ESP): The exchangeable sodium percentage is simply the amount of exchangeable sodium in a soil expressed as a percentage of the total exchangeable ions in the soil. An ESP of more than 15 defines an alkali soil.

ESP = Exchangeable sodium (meq 100 g^{-1} soil)/Cation-exchange capacity (meq 100 g^{-1} soil) $\times$ 100

Flocculation: The joining together of smaller individual particles of soil, especially clay, into larger units or flocs. In soils, presence of calcium tends to flocculate clays, maintain structure, while sodium tends to disperse clays in solution.

Flushing: A process to remove salts from the top few cm by impounding and draining the water through surface drainage. Used for heavy textured or alkali soils where leaching may pose problems. Also see leaching.

Groundwater: Water that occurs beneath the water table in soils and in geologic formations that are fully saturated.

Groundwater quality: Quality of groundwater in relation to purpose for which it is proposed to be used. In the context of this handbook purpose is irrigation.

Gypsum: Gypsum (calcium sulphate) occurs naturally as crystals and is mined. This gypsum is of variable quality and contains impurities. Gypsum is also available as a by-product during phosphate fertilizer production (phosphogypsum), or hydrofluoric acid production (calcium sulphate anhydrite). All of these are used as amendments to reclaim alkali soils.

Gypsum bed: A structure used to dissolve gypsum in irrigation water before it is applied to irrigate the fields.

Gypsum Requirement (GR): The approximate amount of gypsum needed per hectare to lower the ESP of the soil to a desired level. It is calculated as:

GR = (present ESP - desired ESP) $\times$ CEC $\times$ 0.052.

The factor 0.052 assumes CEC is in meq $100g^{-1}$ or cmol (+charge) kg^{-1} units. If CEC is in mmols (+ charge) kg^{-1}, the factor is 0.0052. These factors assume 90% reclamation efficiency and a low SAR of the irrigation water.

Hardpan: A hardened layer of soil (often clay-rich) below the zone of cultivation, caused by cementation (chemical process) and/or extreme compaction (mechanical process) of soil particles. Hardpans may inhibit root growth and trap water, resulting in a waterlogged soil layer.

Humus: Organic matter, also called 'humus', is formed from the decay of leaves, plants and other living organisms.

Hydraulic conductivity: The rate of flow of water through a porous material of a unit cross-sectional area under a unit hydraulic gradient, at a specified temperature. It is also called unit of permeability, transmission constant or coefficient of transmission. It is the coefficient, K in the Darcy's equation $v = -Ki$, such that v is the flow velocity of a specified liquid under saturated conditions, and i is the hydraulic gradient. Thus, unit of K is $L\ T^{-1}$.

Indicator plants: Plants characteristic of specific soil or site condition. The presence of these is used to visually assess the problems.

Infiltration: The process of water entry into the soil, generally (but not necessarily) through the soil surface and vertically downward.

Infiltration rate: The flux density of water passing through the (or a) soil surface and flowing into the soil is known as infiltration rate.

Ion: Ions are electrically charged particles made up of atoms or molecules. Cations are positively charged ions. Anions are negatively charged ions.

Ion exchange: The interchange between an ion in solution and another ion on the surface of any surface-active material such as clay or humus.

Ionic toxicity: The ions, which can be absorbed by the crops in excess quantity to adversely affect the crop. The examples of such ions are: copper, cadmium, boron, arsenic, iron, lead, manganese, etc.

Isobath: An imaginary line or a line on a map or chart that connects all points having the same depth to water table below the soil surface.

Isograms, or Isohaline lines: The locus of points of equal salt concentration.

Land forming: A process of earthmoving to alter the natural topography so as to suit a particular purpose, such as surface irrigation, surface drainage, planting of seeds and rainwater management etc.

Land Reclamation: Changing general physico-chemical characteristics of the land to make it capable of intensive use for agriculture.

Leaching: The process by which soluble materials in the soil, such as salts, nutrients, pesticide chemicals or contaminants, are washed into a lower layer of soil or are dissolved and carried away by water.

Leaching fraction: The fraction of water applied on the soil that actually passes through the root zone as against the leaching requirement which must pass through the root zone to prevent soil salinity from exceeding a specified value.

Leaching requirement: The leaching fraction required to keep soil salinity, chloride, or sodium (whichever is the most limiting factor) from exceeding a tolerance level for the crop rotation. Leaching requirement refers to long-term average conditions.

Lime: Strictly, calcium oxide (CaO), but as commonly used in agricultural terminology, includes calcium carbonate ($CaCO_3$) and calcium hydroxide $Ca(OH)_2$. Agricultural lime refers to any of these compounds, with or without magnesia, used as an amendment for acid soils to make them neutral.

Matric Potential: The energy by which soil water is held by soil particles. This represents the energy required to be expended by plants to take up water from the soil. Matric potential is always negative indicating that energy has to be applied to remove water from the soil.

meq L^{-1}: Milliequivalents per liter, the equivalent of 10^{-3} mol divided by valency.

Mhos: Reciprocal of Ohms. A unit of electrical conductivity to measure the quality of water, See Electrical Conductivity.

Milliequivalents per litre (meq L^{-1}): Equal to ppm /molecular weight.

Milligram per liter (mg L^{-1}): Amount of the substance in mg in one liter of water. It is equal to parts per million (ppm).

Mixing: See blending

Non-saline alkali soil: A soil that contains sufficient exchangeable sodium to interfere with the growth of most crop plants and contains variable quantities of soluble salts. The exchangeable sodium percentage is greater than 15, the conductivity of the saturation extract is usually less than 4 dS m^{-1} (at 25°C) and the pH of the saturated soil paste usually exceeds 8.2.

Organic matter: Any living or dead plant and animal material in the soil is soil organic matter. Organic matter helps to make bonds between soil particles and create and maintain soil structure. It also contains nutrients that are essential for plant growth.

Osmotic head: The head equivalent, in meters, of the osmotic potential or osmotic pressure.

Osmotic Effect: Reduction in the ability of plants to extract water from the soil as the osmotic pressure of the soil water solution increases with increasing salinity.

Osmotic potential (ψo): The osmotic potential (ψo) of the constituent water in soil at a specified temperature is the amount of useful work per unit mass of pure water, in J kg^{-1}, that must be done to transfer reversibly and isothermally an infinitesimal quantity of water from a pool of pure free water to a pool of free soil solution (identical in composition with the soil liquid phase at the point under consideration, and at the same height). Osmotic stress/tension/suction is identical to osmotic pressure. It can be interpreted as the negative of the negative osmotic pressure, in Pascal.

Particle density: Also known as real density or grain density is the average density of the soil particles (solid portion of the soil). It is usually taken as 2.65 g cm^{-3} but may vary from 2.60 g cm^{-3} to 2.75 g cm^{-3}. It is expressed in g cm^{-3}.

Parts per million (ppm): The concentration of a substance expressed as the number of unit weights of that substance per million units of weight of solution. Also expressed as mg kg^{-1} in a solid matrix or mg L^{-1} in solution.

Percolation: A qualitative term applied to the downward movement of water through soil.

Permeability: The capacity or ability of a porous rock, sediment, or soil to allow the movement of water through its pores.

Permeability Index: The permeability Index (PI) for water given by Doneen is calculated by the following equation:

$$PI = 100 \times (Na + \sqrt{HCO_3}) / (Na + Ca + Mg + K).$$

The plot of PI and total salt concentration describes the problems of permeability if any with the continuous use of waters.

pH: See soil pH.

Phosphogypsum: A kind of gypsum available as a by-product of phosphate fertilizer industry. Used as an amendment to reclaim alkali soils and alkali water.

Plant community: An assemblage of plants growing together under similar environmental conditions.

Press mud: A by-product of sugar industry, used in agriculture as a source of nutrient as well as an amendment to reclaim alkali soils if press mud is from industries that use sulphur to clarify sugarcane juice, called sulphitation press mud.

Pressure: Measures force per unit area, Si unit of Pa is equal to 1 N m^{-2}.

Residual Sodium Carbonate (RSC): It is defined as the excess of CO_3+HCO_3 ions over the Ca+Mg ions present in the water. The RSC having units of meq L^{-1} is calculated with the following relation.

$$RSC = (CO_3 + HCO_3) - (Ca + Mg)$$

Resodification: Sodification of reclaimed lands; the process commonly encountered in areas where water table is shallow, irrigation water is alkali and where lands are poorly managed by the farmers.

Saline-sodic soil: A soil containing both sufficient exchangeable sodium and soluble salts to adversely affect the growth of most crop plants. The ESP is >15, the conductivity of the saturation extract > 4 dS m^{-1} (at 25°C) and the pH_s is usually 8.2 or more in the water saturated soil. Also known as saline alkali soils.

Saline soil: A non alkali (sodic) soil containing soluble salts in such quantities that they interfere with the growth of most crop plants. The conductivity of the saturation extract is greater than 4 dS m^{-1} (at 25°C) , the exchangeable sodium percentage is less than 15, and the pH_s of the saturated soil paste is usually less than 8.2.

Saline water: Water that contains significant amounts of dissolved solids and its EC is more than 2.0 dS m^{-1}.

Salinity: Salinity refers to the concentration of all salts in the soil solution. Any readily soluble salt (including fertilizers) may contribute to soil salinity. If the soil solution is saline, plants may have difficulty in taking up both water and nutrients. Some plants may be sensitive to specific ions, such as sodium and chloride and may suffer even in the absence of high salinity.

Salinization: The process of accumulation of salts in the soil particularly in the root zone.

Salt-affected soil: A generic term used to describe soils having an excess of soluble salts, or an excess of exchangeable sodium, or both, such that the growth of most crop plants is adversely affected.

Salt balance: The salt balance of an irrigated area is commonly defined as the difference between the total dissolved solids brought to the land in a pre-decided duration by the irrigation water and the total solids carried away by the drainage water during the same period.

Salt tolerance: The average soil salinity required to produce a specified decrease in plant yield; the ability, expressed qualitatively or quantitatively, of a plant species to withstand salt concentration in soil/water.

Saturated soil paste: A mixture of soil and pure water such that all the voids between the soil particles are filled with water, while at the same time there is no accumulation of free water on the surface. At saturation the soil paste glistens as it reflects light, flows slightly when the container is tipped and the paste slides freely and cleanly off a spatula for all soils, except those with a high clay content.

Saturation extract: The solution obtained by pressure or vacuum filtration of a soil paste that has been made up to a saturated condition by adding water while stirring.

Sediment (sediment): Any material carried in suspension by water, which settles to the bottom if the water velocity is reduced; fine water-borne matter deposited or accumulated in beds. Sediment is ordinarily transported as suspended sediment, by saltation or as bed load.

Sedimentation: The process of subsidence and deposition or settling of suspended matter carried by water or other liquids, by gravity.

Selectivity coefficient: A value describing equilibrium conditions for an exchange reaction. It may be compared with the thermodynamic equilibrium constant, but differs from it in that assumptions are made regarding the activities of the ions participating in the exchange reaction.

Sensitive crops or plants: A qualitative term used to designate the plants which are sensitive to water, salts or some toxic element. Also see tolerant crops.

Soil Shrinkage: Soils that shrink upon drying, a characteristic of the clay minerals of these soils. Often the term include swelling such as shrink-swell soils or shrink-swell capacity of soils.

Siemens: The SI unit for electrical conductance.

Slaked lime: Calcium hydroxide, $Ca(OH)_2$; used in agriculture as a liming material to reclaim acid soils.

Slick spots: Barren areas having puddled or crusted, very smooth, nearly impervious surfaces, usually because of high salinity or sodicity.

Sodication or Sodification: See Alkalization

Sodic soils: A term commonly used synonymously to describe alkali soils. These soils generally have excess of sodium resulting in a dispersed or deflocculated condition. The exchangeable sodium percentage (ESP) is greater than 15. In this condition the water cannot enter the soil profile or is impaired from entering the soil.

Sodic water: Water with residual sodium carbonate more than 2.5 meq L^{-1} and SAR more or less than 15. EC may or may not be high. Sodic water often has more carbonate and bicarbonate ions. Also called alkali water.

Sodification: See alkalization.

Sodium Adsorption Ratio (SAR): A ratio for soil extracts and irrigation water used to express the relative activity of sodium to calcium and magnesium. SAR is calculated as follows:

$$SAR = Na/((Ca+Mg)/2)^{1/2} \text{ (all ions expressed in meq } L^{-1})$$

$$SAR = Na/(Ca+Mg)^{1/2} \text{ (All ions expressed in mmol } L^{-1})$$

It is expressed in units of $(\text{mmol } L^{-1})^{1/2}$.

Soil aeration: The process by which air in the soil is replaced by air from the atmosphere. In a well-aerated soil, the soil air is very similar in composition to the atmosphere above the soil. Poorly aerated soils usually contain a much higher percentage of carbon dioxide and a correspondingly lower percentage of oxygen than the atmosphere above the soil. The rate of aeration depends largely on the volume and continuity of air-filled pores within the soil.

Soil aggregation: The cementing or building together of several soil particles into a secondary unit or ped, aggregate or granule. Water-stable aggregates, which will not disintegrate easily, are of special importance to soil structure.

Soil amendment: See amendment

Soil conditioner: A soil amendment, particularly one that promotes aggregation. Usually refers to the synthetic polymers used for this purpose, e.g. Krilium, PVA, VAMA, HPAN, etc.; Any material added to a soil for the purpose of improving its physical condition.

Soil crust: See crust.

Soil extract: The solution separated from a soil suspension or from a soil by filtration, centrifugation, suction or pressure.

Soil moisture tension: The equivalent negative pressure or suction of water in soil.

Soil pH: Soil pH refers to the acidity or alkalinity of the soil with a pH of 7.0 being neutral. A pH below 7.0 being acid, a pH above 7.0 being alkaline and a pH above 8.5 being alkali because of excess sodium. Most soils with a pH above 7.0 will usually have an excess of undissolved free lime. The optimum pH for most plants is between 6.3 and 7.5, and pH_s below 6.0 require lime for best growth of plants. Soils with pH_s above 8.2 require gypsum for best growth of plants.

Soil porosity: It is defined as the ratio of the volume of voids to the total volume of the soil. Usually denoted by n, it can be determined by the following relation:

$$n = 1 - (BD/PD)$$

Soil separate: Mineral particles < 2.0 mm in diameter, ranging between specified limits. Synonymous with soil particle size class. Refer soil texture.

Soil structure: The combination or arrangement of primary soil particles into secondary units or peds. These secondary units may be, but usually are not, arranged in the profile in such a manner as to give a distinctive characteristic pattern. The secondary units are characterized and classified on the basis of size, shape and degree of distinctness into classes, types and grades, respectively.

Soil texture: The texture is the term indicating the coarseness or fineness of the soil. The amount and quantity of each of the grain size group of particle that constitute the soil. Given by the relative proportion (percent) of the three soil separates (sand, silt, and clay) in a soil.

Soil : water solution at 1:1 or 1:2: One part soil added to one or two part water (weight : weight) to measure pH or electrical conductivity. The measured EC/pH is designated as EC_1/pH_1 or EC_2/pH_2.

Soluble Sodium Percentage (SSP): The proportion of sodium ions in solution in relation to the total cation concentration, defined as follows:

SSP = Soluble sodium concentration (meq L^{-1})/Total cation concentration (meq L^{-1})

The SSP values less than 60 indicates good quality water.

Sorption: General term for the interaction (binding or association) of a solute ion or molecule with a solid

Specific gravity: The density of a material related to that of water whose specific gravity is normally taken as one although exact value depends upon the temperature.

Specific ion effect: Any effect of a salt constituent in the substrate on plant growth which is not due to the osmotic properties of the substrate.

Standard atmosphere: A unit of pressure denoted as atm. It is equal to 1013.25 m bar or 101.325 kPa or 14.696 psi or 760 mm of Hg or 760 torr.

Sub-soil: That portion of the normal soil profile underlying the A horizon. In humid climates it is low in organic matter content, is lighter in colour, usually of finer texture, of higher bulk density, and of lower fertility than the surface soil.

Subsurface drainage: Drainage methods commonly employed to control water table and facilitate leaching of salts e.g. pipe drainage, mole drainage and vertical drainage etc.

Sulphitation press mud: See press mud.

Surface drainage: The diversion or removal of excess water from the surface of the land by means of improved natural or constructed channels, supplemented when necessary by shaping and grading of land surfaces to such channels.

Swelling: The increase in bulk volume of a substance, usually upon addition of water. A swelling soil is one that contains expanding-lattice clay minerals which cause the soil to swell or heave upon wetting and shrink (with the formation of cracks) upon drying.

Synergism: The simultaneous actions of two or more agencies that together have a greater total effect than the sum of their individual effects.

TDS, total dissolved solids: A quantitative measure of the total dissolved organic and inorganic solids concentration in water, an indicator test used for water analysis and also a measure of the mineral content of bottled water and groundwater. One can monitor electrical conductivity to quickly estimate total dissolved solids or TDS. Also TDS is sum of all dissolved materials such as salts, which are non-filterable and remain as evaporate following evaporation of the water.

Tilth: The physical condition of soil as related to its ease of tillage, fitness as a seedbed and its impedance to seedling emergence and root penetration.

Tolerant crops: Capable of withstanding a condition usually un-favourable. Some plants are tolerant to soil salinity and others are tolerant to soil alkalinity. See crop tolerance

Toxic ions: Poisonous; tending to reduce the yield of crops below the normal due to their accumulation in plant leaves to an extent to cause damage to the plant.

Toxic water: Water containing one or more of toxic elements. Yield reduction is attributed to toxic effects of the ions rather than the osmotic potential of the soil solution.

Trace element: An element found in only minor amounts (concentrations less than 1.0 mg L^{-1}) in water or soil or sediment; includes arsenic, cadmium, chromium, copper, lead, mercury, nickel, and zinc

Vertical drainage: Disposal of drainage water through wells into a porous layer of earth or an open-rock formation; the term vertical drainage is also used for a drainage system in which tube wells are used to lower the water table.

Wasteland: Land not suitable to produce material or services of value. A miscellaneous land type.

Water logging: It refers to the condition of land in which it cannot be put to its normal use as a result of either or both of shallow water table and surface ponding. In arable lands depending upon the extent and duration of water logging, the crop yield may reduce substantially or even become zero.

Water quality: A judgment of the chemical, physical and biological characteristics of water and of its suitability for a particular purpose.

Water-stable aggregates: A soil aggregate which is stable to the action of water such as falling rain drops or agitation as in wet sieving analysis.

Water table: The elevation at which the pressure of the water is zero with respect to the atmospheric pressure. The water table is the upper boundary of groundwater.

■■■■

REFERENCES

Abrol I P, and Bhumbla D R. 1971. Saline and alkali soils in India- their occurrence and management. *World Soil Series, FAO Report No. 41*. pp 42-51. Food and Agriculture Organization of the United Nations, Rome.

Abrol I P and Acharya C L. 1975. Soil water behaviour and irrigation frequencies in soils with physical constraints. pp 335-342. *Proc. Second World Congress. International Water Resources Assoc.* 12-16 December. New Delhi.

Abrol I P and Bhumbla D R. 1973. Field studies on salt leaching in a highly saline-sodic soil. *Soil Science* **15**: 429- 33.

Abrol I P, Chhabra R and Gupta R K. 1980. A fresh look at the diagnostic criteria for sodic soils. pp 142-147. *Proc. Int. Symp. on Salt Affected Soils.* 18-21 February, Central Soil Salinity Research Institute, Karnal.

Abrol I P, Dahiya I S and Bhumbla D R. 1975. On the method of determining gypsum requirement of soils. Soil Science **120**: 30-6.

Abrol I P, Dargan K S and Bhumbla D R. 1973. *Reclaiming Alkali Soils*. Bull. **2**. Pages 58. Central Soil Salinity Research Institute, Karnal.

Abrol I P Gupta R K and Singh S B. 1979. Note on solubility of gypsum and sodic soil reclamation. *J. Ind. Soc. Soil Science* **27**: 482–83.

Abrol I P, Saha A K and Acharya C L. 1978. Effect of exchangeable sodium on some physical properties. J. Indian Society Soil Science **26**: 98-105.

Abrol I P, Yadav J S P and Massoud F I. 1988. *Salt Affected Soils and Their Management*. FAO Soils Bulletin 39. Food and Agriculture Organization of the United Nations, Rome.

Acharya C L and Abrol I P. 1975. A comparative study of soil water behavior in a sodic and an adjacent normal soil. *J. Indian Soc. Soil. Science* **23**: 391-401.

Acharya C L and Abrol I P. 1976. Effect of river sand on the permeability of a sodic soil. *J. Indian Soc. Soil Science* **24**: 245-52.

Acharya C L and Abrol I P. 1991. *Soil Water Relations and Root Water Extraction in Alkali Soils*. Tech. Bull. **16**. Pages 96. Central Soil Salinity Research Institute, Karnal.

Agarwal R R and Gupta R N. 1968. *Saline and Alkali Soils in India*. Bull. **15**. Pages 221. Indian Council of Agricultural Research, New Delhi.

Agarwal R R, Yadav J S P and Gupta R N. 1982. *Saline and Alkali Soils of India*. Pages 285. Indian Council of Agricultural Research, New Delhi.

Agboma P, Jones M G K, Peltonen-Sainio P, Hinkkanen R and Pehu E. 1997. Exogenous glycine betaine enhances grain yield of maize, sorghum and wheat grown under two supplementary watering regimes. *Journal of Agronomy and Crop Science* **178**: 29–37.

Allison L E and Moodie C D. 1965. Carbonate. pp. 1379-96. *Methods of Soil Analysis* Part 2. Black C A (Ed.). American Society of Agronomy, Madison.

Anonymous. 2007. *Biennial Report (2004-06)*. Pages 171. All India Coordinated Research Project on Management of Salt Affected Soils and Use of Saline Water in Agriculture. Central Soil Salinity Research Institute, Karnal.

Anonymous. 2009. *Biennial Report (2007-08)*. Pages 212. AICRP on Management of Salt Affected Soils and Use of Saline Water in Agriculture. Central Soil Salinity Research Institute, Karnal.

ANZECC (Australian and New Zealand Environment and Conservation Council). 2000. *Australian and New Zealand Guidelines for Fresh and Marine Water Quality: The Guidelines*. Pages 314. Paper No. 4. Agriculture and Resource Management Council of Australia and New Zealand, Canberra.

Arora S, Gururajarao G and Sharma D K. 2015. Exploiting plant-microbe interactions for remediation of salt affected soils. *Salinity News* **21**. 4.

Arora S, Singh Y P, Vanza M and Sahni, D. 2016. Bio-remediation of saline and sodic soils through halophilic bacteria to enhance agricultural production. *Journal of Soil and Water Conservation* **15**: 302-05.

Arora S and Mehta R. 2018. Characteristics of coastal saline soils inhabiting halophytes and salt tolerant plants. *Journal of Soil and Water Conservation* **17**: 107-10.

Ashraf M and Foolad M R. 2007. Roles of glycine betaine and proline in improving plant abiotic stress resistance. *Environmental and Experimental Botany* **59**: 206–16.

Ashworth J, Keyes D, Kirk R and Lessard R. 2001. Standard procedure in the hydrometer method for particle size analysis. *Communications in Soil Science and Plant Analysis* **32**: 633-42.

Ayers R S and Westcot D W. 1976. *Water Quality for Agriculture*. Irrigation and Drainage Paper, **29**. Pages 97. Food and Agriculture Organization of the United Nations, Rome.

Ayers R S and Westcot D W. 1985. *Water Quality for Agriculture*. Irrigation and Drainage Paper, **29**, Revision- 1. Pages 109. Food and Agriculture Organization of the United Nations, Rome.

Bains S S and Singh K N. 1966. Utilization of solar radiation in desalinization of ridged plant beds on saline soil. *Nature* **212**: 1391-92.

Bal A R. 1976. Salinity tolerance through seed treatment with proline. *Biol. Pl.* **18**: 227-29.

Bandyopadhyay K K, Aggarwal P, Chakraborty D, Pradhan S, Garg R N and Singh R. 2012. *Practical Manual on Measurement of Soil Physical Properties*. Pages 62. Division of Agricultural Physics, Indian Agricultural Research Institute, New Delhi.

Batra L, Kumar A and Manna, M C. 1997.Microbiological and chemical amelioration of alkaline soils by growing Karnal grass and gypsum application. *Experimental Agriculture* **33**: 389-94.

Beek C G E M van and Breemen N van 1973. The alkalinity of alkali soils. *Journal Soil Science* **24**: 129-36.

Beyce O. 1972. *Drainage of Salty Soils*. Irrigation and Drainage Paper **16**. pp 63-67. Food and Agriculture Organization of the United Nations, Rome.

Bhargava G P. 2003. *Training Manual for Undertaking Studies on Genesis of Sodic/Alkali Soils*. Pages111. Central Soil Salinity Research Institute, Karnal.

Bhargava, G.P., Sharma, R.C., Pal, D.K. and Abrol, I.P. 1980. A case study of the distribution of salt affected soils in Haryana state. Int. Symp. Salt Affected Soils. CSSRI, Karnal. 83-91.

Bhargava G P, Singhla S K and Gupta I C. 1974. Distribution of boron in a salt affected and contiguous normal soils. *Ann. Arid Zone* **13**: 32-8.

Bhumbla D R and Abrol I P. 1975. Effect of application of different levels of gypsum on the yield of rice, wheat and barley grown in a saline soil. pp 537-544. *Proc. Int. Symp. New Development in the Field of Salt Affected Soils,* 4-9 December, Cairo.

Bhumbla D R and Dhingra D R. 1964.Micro-nutrient status of saline and alkali soils of the Punjab. *J. Indian Soc. Soil Science* **12**: 255-60.

Bisutti I, Hike I and Raessler M. 2004. Determination of total organic carbon- an overview of current methods. *Trends in Analytical Chemistry* **23**: 716-26.

Borojevic S, Cupina T and Krsmanovic M. 1980. Green area parameters in relation to grain yield of different wheat genotypes. *Zeitschrift fu¨r Pflanzenzu¨chtung* **84**: 265–83.

Bos M G. 1977. Arid land irrigation in developing countries- environmental problems and effects. pp 193-204. Worthington E. B. (Ed.), Pergamon Press, Oxford.

Bower C A. 1959. Cation exchange equilibrium in soils affected by sodium salts. *Soil. Science* **88**: 32-5.

Chakravarti P and Chakravarti S. 1957. Soils of West Bengal. *Annals of Indian National Science Academy* **23**: 93-101.

Chawla K L. 1985. *Salt and Water Transport under High Water Table Conditions.* Unpublished Ph. D. thesis. Kurukshetra University, Kurukshetra.

Chawla K L and Abrol I P. 1982. Effect of gypsum fineness on the reclamation of sodic soils. *Agricultural Water Management* **5** : 41-50.

Cheng K and Bray R H. 1951. Determination of calcium and magnesium in soil and plant material. *Soil Science* 72: 449-58.

Chepil W S. 1962. A compact rotary sieve and the importance of dry sieving in physical soil analyses. *Soil Sci. Soc. Am. Proceedings* **26**: 4-6.

Chhabra R. 2004. Classification of Salt-Affected Soils. *Arid Land Research and Management* **1**: 61-79.

Chhabra R and Abrol I P. 1977. Reclaiming effect of rice grown in sodic soils. *Soil Science* **124**: 49-55.

Chopra S L and Kanwar J S. 1991. *Analytical Agricultural Chemistry*. Kalyani Publishers, New Delhi.

Choudhary O P. 2017. Long-term impact of cyclic use of sodic and canal waters for irrigation on soil properties and crop yields in cotton-wheat rotation in a semiarid climate. *Agricultural Research* **6**: 267-72.

Craze B and Hamilton G J. 1991. Soil physical properties. In: Soils: Their Properties and Management: A Soil Conservation Handbook for New South Wales. Charman P E V and Murphy B W (Eds) Sydney University Press, Sydney.

CSSRI 1997. *Vision-2020. CSSRI Perspective Plan.* Pages 95. Indian Council of Agricultural Research. Central Soil Salinity Research Institute. Karnal, India.

CSSRI. 2007. *CSSRI- Perspective Plan: Vision 2025.* Pages 127. Central Soil Salinity Research Institute, Karnal.

CSSRI. 2011. *CSSRI Vision 2030.* Pages 26. ICAR- Central Soil Salinity Research Institute, Karnal.

CSSRI. 2015. *CSSRI Vision 2050.* Pages 31. ICAR- Central Soil Salinity Research Institute, Karnal.

Dagar J C, Singh G and Singh N T 2001. Evaluation of forest and fruit trees used for rehabilitation of semiarid alkali-sodic soils in India. *Arid Land Research and Management* 15:115-33.

Dahiya I S, Malik R S and Singh M. 1981. Field studies on leaching behaviour of highly saline-sodic soil under two modes of water application in presence of crops. *J. Agric. Sci. Cambridge* **97**:383-89.

Dahiya I S, Malik R S and Singh M. 1982. Reclaiming a saline-sodic, sandy loam soil under rice production. *Agricultural Water Management* 5: 61-72.

Dahiya I S, Singh M, Richter J and Singh M. 1984. Leaching of soluble salt during infiltration and redistribution. *Irrigation Science* 5: 15-24.

Damodaran T, Sharma D K, Mishra V K and Jha S K. 2014. Rhizosphere engineering: an innovative approach for sustainable crop production in sodic soils. In: *4th National Seminar on Innovative Saline Agriculture in Changing Environment.* 12-14 December, Gwalior.

Dandekar C B and Chougule B A. 2010. Drainage of irrigated lands. pp 1-11. *Proc. 9th Int. Drainage Symposium*, CIGR and CSBE, 13-16 June. Canada.

Dargan K S and Chillar R K. 1973. Cultivation technique for sugar beet and vegetables in saline alkali soils. *Indian Farming* **22**: 13-6.

Dargan K S, Gaul B L and Abrol I P. 1973. Growing sugarcane in saline-alkali soils. *Indian Farming* **23**: 10, 19.

Datta S, Thibault Y. Fyfe W S, Powell M A, Hart B R, Martin R R and Tripthy S. 2002. Occurrence of trona in alkaline soils of the Indo-Gangetic Plains of Uttar Pradesh (UP), India. *Episodes* **25**: 236-39.

Deivanai S, Xavier R, Vinod V, Timalata K and Lim O F. 2011. Role of exogenous proline in ameliorating salt stress at early stage in two rice cultivars. *J. Stress Physiol. Biochemistry* **7**: 157–74.

De'Sigmond A A J. 1925. *Contribution to the Theory of the Origin of Alkali Soils.* (Translated Waksman, S.A.). U. S .Department of Agriculture, Washington, D. C.

Dhir R P. 1998. Origin and accumulation of salts in soils and ground waters. pp 41-60. In: *Agricultural Salinity Management in India.* Tyagi, N K and Minhas P S (Eds.). Central Soil Salinity Research Institute, Karnal.

Diserens E. 1934. Beitrag zur Bestimmung der Durchlässigkeit des Bodens in natürlicher Bodenlagerung. Pages 12. Schweiz. Landw. Monatstr.

Doneen L D. 1963. *Water Quality for Agriculture.* Pages 48. Department of Irrigation, University of California. Davis.

Dwivedi R S. 1979. *Plant Physiology.* A Decade of Research. pp 99-116. Central Soil Salinity Research Institute, Karnal.

Eaton F M. 1950. Significance of carbonates in irrigation waters. *Soil Science* **69**: 123-33.

El-Gabaly M M. 1971. Reclamation and management of salt affected soils. pp 50-79. *FAO Irrigation and Drainage Paper 7*. Food and Agriculture Organization of the United Nations, Rome.

Ernst L F. 1950. A new formula for the calculation of the permeability factor with the auger holes. *Soil. Sc. Soc. Am. J.* **15**: 90-6.

Gapon E N. 1933.Theory of exchange adsorption in soils. *J. Gen. Chemistry USSR* **3**: 144-52.

Garrels R M and Christ C L. 1965. *Solutions, Minerals and Equilibria*. Pages 450. Harper and Row, New York.

Gupta I.C. 1972. Note on relationship between pH of saturated soil paste and 1:2 soil water suspension. *Indian J. Agric. Science* **42**: 962-63.

Gupta I C and Abichandani C T. 1970. Seasonal variations in salt composition of some saline water irrigated soils of western Rajasthan I. Effect of rainfall. *J. Indian Soc. Soil. Science* **18**: 428-35.

Gupta I C and Chandra H. 1972. Effect of gypsum in reducing boron hazard of saline irrigation waters and irrigated soil. *Annals of Arid Zone* 11: 228-30.

Gupta I C and Gupta S K. 2019. *Crop Production in Salt Affected Soils*. 2nd Revised Edition. Pages 412. Scientific Publishers (India), Jodhpur.

Gupta I C, Yaduvanshi N P S and Gupta S K. 2012. *Standard Methods for Analysis of Soil Plant and Water*. Pages 224. Scientific Publishers (India), Jodhpur.

Gupta R N and Narain B. 1971. Investigations on some physical properties of alluvial soils of Uttar Pradesh related to conservation and management. *J. Indian Soc. Soil Science* **19**: 1-9.

Gupta S K. 1981. Evaluation of salt and water balance in saline soils without provision of drains. pp 112-14. *Annual Report-1981*. Central Soil Salinity Research Institute, Karnal.

Gupta S K. 1985. Subsurface drainage for waterlogged saline soils. *Irrigation and Power Journal* **42**: 335-44.

Gupta S K. 1985. Dynamics of salts in saline water irrigated soils. *J. Institution of Engineers (India)* **41**: 17-22.

Gupta S K. 1992. *Leaching of Salt Affected Soils*. Technical Bulletin No. 17. Pages 87. Central Soil Salinity Research Institute, Karnal.

Gupta S K. 2010. *Management of Alkali Water*. Technical bulletin. Pages 62. Central Soil Salinity Research Institute, Karnal, India.

Gupta S K. 2011. Case studies on subsurface drainage systems and guidelines for design and evaluation. pp 11-22. *Salinity Management for Sustainable Agriculture in Canal Commands*. Dey P and Gupta, S K (Eds.). Central Soil Salinity Research Institute, Karnal.

Gupta S K. 2015. Assessing the hazards of high SAR and alkali water: a critical review. *J. Soil Salinity and Water Quality* **7**: 1-11.

Gupta S K. 2019. *Drainage Engineering: Principles and Practices*. Pages 531. Scientific Publishers (India) Jodhpur.

Gupta S K and Gupta I C. 2014. *Management of Saline and Waste Water in Agriculture*. Pagres 315. Scientific Publishers (India) Jodhpur.

Gupta S K and Pandey R S. 1983. A mathematical model for evaluating leaching of saline soils when saline and fresh water are used in conjunction. *Indian J. Agric. Science* 53: 701-06.

Gupta S K, Sharma D P, Tyagi N K and Dubey S K. 2002. *Lavaniya-Kshariya Mirdaon ka Sudhar aur Parbandh*. Pages 153. Central Soil Salinity Research Institute, Karnal.

Gupta S K and Sharma S K. 1990. Response of crops to high exchangeable sodium percentage. *Irrigation science* 11:173-79.

Gururaja Rao G, Nayak A K, Chinchmalatpure A R, Singh R and Tyagi N K. 2001. *Resource Characterization and Management Options for Salt Affected Black Soils of Agro-Ecological Region V of Gujarat State*. Technical Bulletin **11/2001.** Pages 83., Central Soil Salinity Research Institute, Regional Research Station, Anand.

Hardy L A and Eugster H P. 1970. The Evolution of Closed Basin Brines. *Mineral. Soc. Am. Special Publication* **3**: 273-90.

Hilgard E W. 1886. *Alkali Lands, Irrigation and Drainage in Their Mutual Relations***.** Pages 56. Calif. Agr. Expr. Sta. Ann. Rept. California, USA.

Hilgard E W. 1906. Soils, Their Formation, Properties, Composition, and Relations to Climate and Plant Growth. Pages 484. The Macmillan Co., New York.

Hillel D. 1980. *Fundamentals of Soil Physics*. Pages 413. Academic Press, New York, NY.

Hira G S, Bajwa M S and Singh N T. 1981. Prediction of water requirement for gypsum dissolution in sodic soils. *Soil Science* **131**: 353-58.

Hira G S and Singh N T. 1978. Irrigation water requirement for dissolution of gypsum in sodic soil. *Soil Science Soc. Am. Journal* **44**: 930-33.

Hoffman G J. 1980. Guidelines for reclamation of salt affected soils. pp 49-64. *Proc. Inter-American Salinity and Water Management*. Tech. Conf. Juarez, Mexico.

Hoffman G J and van Genuchten M Th. 1983. Soil properties and efficient water use: water management for salinity control. pp 73-85. *Limitations of Efficient Water Use in Crop Production*. Am. Soc. Agronomy, Madison, USA.

Hooghoudt S B. 1936. Bepailing van den doorlaatfactor van den grond met behulp van pompproeven, (z.g. boorgatenmethode). *Verslag Landbouwk. Onder zoek* **42**: 449-541.

Hoon R C and Mehta M L. 1937. *A Study of the Soil Profile of the Punjab Plains with reference to their Natural Flora*. Publication No. **3**. Punjab lIrrigation Research Institute, Amritsar.

http://www.fao.org/soils-portal/soil-management/management-of-some-problem-soils/salt-affected-soils/more-information-on-salt-affected-soils/en/. Opened on December 12,2018.

Jackson M L. 1973. *Soil Chemical Analysis*. Pages 498. Prentice Hall of India Pvt. Ltd., New Delhi.

Jagtap T G, Bhosale S H and Nagle V L. 2002. Ecological observations on major *Salicornia* beds from highly saline coastal wetlands of India. *Wetlands* **22**: 443–50.

Jain T C, Saxena G S, Jain K C and Totawat K L. 1976. Effect of crop rotation, farm yard manure and the quantity of saline irrigation water on the yield of wheat and physico-chemical properties of the soil. *Indian J. Agric. Research* **10**: 1-9.

Joshi Y C, Qadir Ali, Bal A R and Rana R S. 1980. Sodium/potassium index of wheat seedling in relation to sodicity tolerance. pp 457-460. *Proc. Int. Symp. on Salt Affected Soils*. 18-21 February, Central Soil Salinity Research Institute, Karnal.

Jury W A, Jarrell W M and Devitt D. 1979. Reclamation of saline-sodic soil by leaching. *Soil Sci. Soc. Am. J.* **43**: 1100- 06.

Kinney J L. 1979. *Laboratory Procedures for Determining the Dispersibility of Clayey Soils*. Report No. **REC-ERC-79-1O**. Pages 27.Bureau of Reclamation, Denver, CO.

Kittrick J A. 1971.Soil Solution composition and stability of clay minerals. *Soil Sci. Soc. Am. Proceedings* **35**: 450-54.

Klute A and Dirksen C. 2006. Hydraulic conductivity and diffusivity: laboratory methods. pp 687-732. *Methods of Soil Analysis. Part 1.* Agron. Monogr. **9**, 2nd ed. Klute, A. (Ed.). ASA and SSSA, Madison, WI.

Knodel P C. 1991. *Characteristics of Dispersive and Problems Clay Soils*. Bureau of Reclamation Denver Office. **R91-09**. Pages 18. U.S. Department of the Interior. Denver, Colorado.

Konen M E, Jacobs P M C, Lee B B, Talaga J and Mason J A. 2002. Equations for predicting soil organic carbon using loss-on-ignition for North Central US soils. *Soil Science Society of America Journal* **66**: 1878-81.

Kostiakov A N. 1932. On the dynamics of the coefficient of water percolation in soils and on the necessity of studying it from a dynamic point of view for purposes of amelioration. *Trans. Com. Int. Soc. Soil. Sci. Part A*, 6th Ed. Moscow. 17-21.

Kovda V A, Berg C V D and Hagan R M. 1973. *Irrigation, Drainage and Salinity: An International Source Book*. FAO/UNESCO. Hutchison and Co. London.

Krishnamurthy S L, Pundir P, Singh Y P, Sharma S K, Sharma P C and Sharma D K. 2015. Yield stability of rice lines for salt tolerance using additive main effects and multiplicative interaction analysis – AMMI. *Journal of Soil Salinity and Water Quality* **7**: 98-106.

Krishnamurthy S L, Sharma P C, Sharma D K, Ravi Kiran K T, Singh Y P, Mishra V K, Burman D, Maji B, Mandal S, Sarangi S K, Gautam R K, Singh P K, Manohara K K, Marandi B C, Padmavathi G, Vanve P B, Patil K D, Thirumeni S, Verma O P, Khan A H, Tiwari S, Geetha S, Shakila M, Gill R, Yadav V K, Roy S K B, Prakash M, Bonifacio J, Ismail A, Gregorio G B and Singh R K. 2017. Identification of mega-environments and rice genotypes for general and specific adaptation to saline and alkaline stresses in India. *Scientific Reports* **7**: 7968 (DOI: 10.1038/s41598-017-08532-7)

Kumar A and Abrol I P. 1986. *Grasses in Alkali Soils*. Bulletin No. **11**. Pages 95. Central Soil Salinity Research Institute, Karnal.

Kumar A, Lata C, Krishnamurthy S L, Kumar A, Prasad K K R and Kulshreshtha N. 2017. Physiological and biochemical characterization of rice varieties under salt and drought stresses. *Journal of Soil Salinity and Water Quality* **9**: 167-77.

Kumar M, Hasan M, Arora A, Gaikwad K, Kumar S, Rai R D and Singh A. 2015. Sodium chloride-induced spatial and temporal manifestation in membrane stability index and protein profiles of contrasting wheat (*Triticum aestivum* L.) genotypes under salt stress. *Indian J. Plant Physiology* **20**: 271–75.

Kumar, R. 1992. *Genesis of Alkali Soils at Different Geomorphic Locations in Haryana*. Ph.D. Thesis, Haryana Aagricultural University, Hisar.

Kumar R, Ahuja R L, Singh N T and Ghabru S K. 1994. Characteristics and classification of alkali soils on Siwalik hills of Satluj-Yamuna divide, north-west India. *Agropedology* **5**: 29-35.

Kumar R, Ghabru S K, Ahuja R L and Singh N T. 1993. Elemental composition and mineralogy of alkali soils on the Siwalik Hills. *Agropedology* **4**: 59-68.

Lefellar P A and Sharma R P. 1977. Leaching of a highly saline-sodic soil. *J. of Hydrology* **32**: 203-18.

Lindsay W L. 1979. *Chemical Equilibriums in Soils*. Pages 449. John Wiley &Sons, New York.

Lutts S. 2000. Exogenous glycine betaine reduces sodium accumulation in salt stressed rice plants. *International Rice Research Notes* **25**: 39–40.

Maas E V. 1980. Crop tolerance to saline soil and water. pp 205- 211. *Prospectus for Bio-saline Research.* Proc. US –Pak Bio-saline Res. Workshop, Karachi.

Maas E V. 1984. Salt tolerance of plants. *The Handbook of Plant Science in Agriculture.* Christie B R (Ed). CRC Press, Boca Raton, Florida, USA.

Maas E V. 1986. Salt Tolerance of plants. *App. Agric. Research* **1**: 12-26.

Maas E V. 1987. Salt tolerance of plants. *Handbook of Plant Science in Agriculture*, Vol. 2. Christie B R (Ed.). CRC Press, Boca Raton, FL, USA.

Maas E V and Hoffman G J. 1977. Crop salt tolerance: current assessment. *J. Irrig. Drainage Division* **103**: 115- 34.

Mehta M L. 1951. *Land Reclamation*. Pages 117. Central Board Irrigation Power, New Delhi.

Mendpara J, Parekh V, Vaghela1 S, Makasana A, Kunjadia P D, Sanghvi G, Vaishnav D and Dave G S. 2013. Isolation and characterization of high salt tolerant bacteria from agricultural soil. *European Journal of Experimental Biology* **3**:351-58.

Minhas P S. and Gupta R K. 1992. *Quality of Irrigation Water – Assessment and Management*. Pages 123. Indian Council of Agricultural Research, New Delhi.

Minhas P S and Sharma D R. 2006. Predictability of existing indices and an alternative coefficient for estimating sodicity build-up using adj R_{Na} and permissible limits for crops grown on soils irrigated with waters having residual alkalinity. *J. Ind. Soc. Soil Science* **54**: 331-38.

Ministry of Agriculture. *Wastelands of India- 1984-85*. Govt. of India, Publication Division, New Delhi.

Ministry of Agriculture. 1994. *Draft Report on Status of Land Degradation in India.* Pages 102. Min. of Agriculture. New Delhi. 102 p.

Misra B. 1996. *Highlights of Research on Crops and Varieties for Salt Affected Soils.* Research Bulletin. Pages 28. Central Soil Salinity Research Institute, Karnal.

Mondal A K, Obi Reddy G P and Ravisankar T. 2011. Digital database of salt affected soils in India using Geographic Information System. *Journal of Soil Salinity and Water Quality* **3**:16-29.

Muhr G R, Datta N P, Sankarasubramoney H, Dever R F, Leley V K and Donahue R L. 1963. *Soil Testing in India*. Pages 49. United States Agency for International Development Mission of India. New Delhi.

Nakayama F S and Bucks D A. 1991. Water quality in drip/trickle irrigation: a review. *Journal of Irrigation Science* **12**: 187–92.

National Commission on Agriculture (NCA). 1976. *Report of the Commission, Climate and Agriculture (Part iv).* Ministry of Agriculture and Irrigation. Govt. of India, New Delhi.

Nayak A K, Sharma D K, Mishra V K, Jha S K, Singh C S, Shahabuddin Md, Srivastava M, Kale S P and Singh G B. 2011. *Phospho-gypsum: An Alternate Chemical Amendment for Sodic Lands.* Tech. Bull.**1/2011**. Pages 42. Central Soil Salinity Research Institute, RRS, Lukhow.

Nayak A K, Sharma D K, Mishra V K, Minhas P S and Verma C L. 2008. Reclamation of saline sodic soil under a rice wheat system by horizontal surface flushing. *Soil Use and Management*. **24**: 237-43.

Neelakanthan S and Pahwa M R. 1977. Note on the distribution of micro-organisms in normal, saline and alkali soils. *Current Agriculture* **1**: 77-8.

Nielsen D R and Bigger J W. 1961. Miscible displacement in soils. *Soil Science Society America Proceeding* **25**: 1-5.

Nielsen D R, Biggar J W and Luthin J N. 1966. Desalinization of soils under controlled unsaturated flow conditions. pp 15-24. *6th Int. Cong. Irrigation and Drainage.* New Delhi.

Oster J D. 1982. Gypsum usage in irrigated agriculture: A review. *Fertilizer Research* 3:73-89.

Oster J D. and Halvorson A D. 1978. Saline seep chemistry. pp 2.7–2.29. *Dry Land - Saline - Seep Control.* 11th International Soil Science Soc. Cong., Edmoton.

Overstreet R, Martin J C and King H M. 1951. Gypsum, sulfur, and sulfuric acid for reclaiming an alkali soil of the Fresno series. *Hilgardia* **21**: 113-41.

Paikray S P and Singh K. 2012. Microbial reclamation of saline and sodic soil. *WO 2012046098* http://www.google.com/patents/ WO2012046098A1.

Pal D K, Dasog G S, Vadivelu S, Ahuja R L and Bhattacharyya T. 2000. Secondary calcium carbonate in soils of arid and semi-arid regions of India. pp 149-185. *Global Change and Pedogenic Carbonates.* Lal R, Kimble J M and Stewart B A (Eds.). Lewis Publishers, Boca Raton Florida USA.

Paliwal K V. 1967. A comparison of some of the ion exchange equations. *Proceedings of the Indian National Science Academy* **33**: 231-48.

Paliwal K V and Maliwal G L. 1971. Prediction of exchangeable sodium percentage from cation exchange equilibria. *Geoderma* **6**: 75-8.

Pearson G A 1960. *Tolerance of Crops to Exchangeable Sodium.* Bull. **216**. Pages 4. US Dept. Agric. Information Washigton.

Piper C S. 1942. *Soil and Plant Analysis.* Pages 368. The University of Adelaide, Adelaide.

Poonia S R, Jhorar L R, Nath J and Khanna S S. 1974. Effect of quality of irrigation water, leaching level and farm yard manure on performance of wheat and pearl millet. *Indian J. Agric. Science* **44**: 854-59.

Puntamskar S S, Mehta P C and Seth S P. 1971. Note on the inducement of salt resistance in two wheat varieties by presoaking with different salts of varying concentration. *Indian j. agric. Science* **41**: 717-18.

Puri A N. 1930. *A New Method for Estimating Total Carbonates in Soils.* Bull. No. **206**. Pages 7. Imp. Agr. Res. Institute Pusa.

Quershi R H and Barret-Lennard E G. 1998. *Saline Agriculture for Irrigated Lands in Pakistan: A Handbook.* Monograph **50**. Pages 142. Australian Centre for International Agricultural Research, Canberra, Australia.

Quirk J P and Schofield R K. 1955. The effect of electrolyte concentration on soil permeability. *J. Soil Science* **6**: 163-78.

RAJAD, 1995. *Analysis of Subsurface Drainage Criteria.* Rajasthan Agricultural Drainage Research Project, Kota (Rajasthan).

Ranade D H. 1984. *Leaching Phenomenon in Sandy Loam Soils.* Unpublished M. Tech. Thesis, Indian Institute of Technology, Kharagpur.

Ranade D H and Gupta R K. 1987. Salt effect on hydraulic conductivity of a Vertisol. *Catena.* **14**: 45-54.

Rao K V G K, Singh O P, Gupta R K, Kamra S K, Pandey R S, Kumbhare P S and Abrol I P. 1986. *Drainage Investigations for Salinity Control in Haryana.* Bull. No. **10**. pages 95. Central Soil Salinity Research Institute, Karnal.

Reeve R C, Pillsbury A F and Wilcox L V. 1955. Reclamation of a saline and high boron soil in the Coachella Valley of California. *Hilgardia* **24**: 69-91.

Reeve R C and Doering E J. 1966. The high-salt-water dilution method for reclaiming sodic soils. *Soil Sci. Soc. Amer. Proceeding* **30**: 498-504.

Rengasamy P. 2006. World salinization with emphasis on Australia. *Journal of Experimental Botany* **57**: 1017-23.

Rengasamy P. 2010. Soil processes affecting crop production in salt-affected soils. *Functional Plant Biology* **37**: 613–20

Report of the Working Group. 1991. *Problem Identification in Irrigated Areas with Suggested Remedial Measures.* Ministry of Water Resources, Government of India, New Delhi.

Rhoades J D. 1972. Quality of water for irrigation. *Soil Science* **113**: 277-84.

Rhoades J D. 1983. Using saline water for irrigation. pp 233-64. Proc. Intern. Workshop Salt Affected Soils of Latin America. 23-30 October. Maracay Venezuela.

Richards L A (Ed.). 1954. *Diagnosis and Improvement of Saline and Alkali Soils.* Handbook No . **60**. Pages 160. US. Dept. of Agriculture. US Salinity Laboratory, Riverside CA.

Ritzema H P. 1994. *Drainage Principles and Applications.* ILRI Pub. **16**. Pages 1125. International Institute for Land Reclamation and Management (ILRI), Wageningen.

Sandhu S S, Acharya C L and Abrol I P. 1980. The effect of exchangeable sodium on the water retention, transmission and modulus of rupture of a soil. pp 185-94. *Proc. Int. Symp. on Salt Affected Soils.* 18-21 February, Central Soil Salinity Research Institute, Karnal.

Santhosh Kumar M and Thiyageshwari S. 2018. Performance of nano-gypsum on reclamation of sodic soil. *International Journal of Current Microbiology and Applied Sciences* **6**: 56-62.

Satyanarayana K V S. 1961. Studies on the soils of Rann of Kutch I: Description of some profiles and distribution of salts and gypsum in the profiles. *Bull. Indian Soc. Soil Science* **6**: 125-33.

Saxena P N. 1960. Algal flora of *usar* lands in Uttar Pradesh. *Sci. Culture* **25**:542-43.

Schoonover W R. 1952. *Examination of Soils for Alkali*. Univ. of California. Extension Service, Berkley, California.

Sehgal J L. and Stoop G. 1972. Pedogenic calcite accumulation in arid and semi-arid regions of the Indo-Gangetic plains of Punjab. *Geoderma* **8**: 59-72.

Sharma D P and Gupta S K. 2006. Subsurface drainage for reversing degradation of waterlogged saline lands. *Land Degradation and Development* **17**: 605-14.

Sharma P and Swarup A. 1997. Comparison of pyrites varying in water-soluble sulfur with gypsum for the reclamation of alkali soils under a rice-wheat rotation. *Biol. Fertil. Soils* **24**: 96-101.

Sharma P C. 2003. Salt tolerance of Indian mustard (*Brassica juncea* L.): factors affecting growth and yield. *Indian Journal of Plant Physiology* **8**: 368-72.

Sharma P C and Gill K S. 1994. Salinity induced effects on biomass, yield, yield attributing characters and ionic contents in genotypes of Indian mustard (*Brassica juncea*). *Indian Journal of Agricultural Sciences* **64**: 785-88.

Sharma P C. and Kumar P. 1999. Alleviation of salinity stress during germination in *Brassica juncea* by pre-sowing chilling treatments to seeds. *Biologia Plantarum* **42**: 451- 55.

Sharma R C and Bhargava G P. 1981. Morphogenic changes in an alkali soil pedon during amelioration through gypsum application. *J. Ind. Soc. Soil Science* **29**: 274-75.

Sharma S K. 2000. Ground water management. *Proc. Conf. 3rd Water Asia 2000*. Interads Ltd. New Delhi.

Sigmond A A J. 1927. *De-Hungarian Alkali Soils and Methods of Their Reclamation* (English translation published in 1927). Pages 156. Calif. Agr .Expr. Sta., University of California, USA.

Singh A K and Verma S K. 2016. Hydro-physical properties and solute movement in black alkali soils. Pp 119-144. *Innovative Saline Agriculture* Dagar J C, Sharma P C, Sharma, D K and Singh A K. (Eds). Springer (India) Pvt. Ltd. New Delhi.

Singh G B and Dagar J C. 2005. *Greening Sodic Lands: Bichhian Model*. Tech. Bull. **2/2005**. Pages 51. Central Soil Salinity Research Institute, Karnal.

Singh G B and Singh N T. 1993. *Mesquite for the Revegetation of Salt Lands*. Bull. **18**. Pages 24. Central Soil Salinity Research Institute, Karnal.

Singh G B, Singh N T and Tomar O S. 1993. *Agroforestry in Salt-affected Soils*. Technical bulletin No.**17.** Pages 65. Central Soil Salinity Research Institute, Karnal.

Singh J, Singh V, Vineeth T V, Kumar P, Kumar N and Sharma P C. 2018. Differential response of Indian Mustard (*Brassica juncea* L., Czern & Coss) under salinity: photosynthetic traits and gene expression. *Physiology and Molecular Biology of Plants* (Published online 12 Dec 2018, DOI: 10.1007/s12298-018-0631-3)

Singh K N and Sharma P C (Eds.) 2005. *Crop Improvement for Management of Salt Affected Soils*. Pages 221. Central Soil Salinity Research Institute, Karnal.

Singh O P. 1980. Analysis of the hydrologic changes taking place in alkali soil watersheds during reclamation. Ph.D. Thesis JNTU, Hyderabad.

Singh R N. 1950. Reclamation of *usar* lands in India through blue green algae. *Nature (London)* **165**: 325-26.

Singh N T. 1992. Salt affected soils in India. pp 65-100. *Land and Soils*. Khoshoo T N and Deekshatulu B L (Eds.). Indian National Science Academy, New Delhi.

Singh N T. 1974. Physico-chemical changes in sodic soils incubated at saturation. *Plant Soil* **40**: 303-11.

Sinha T S, Singh D, Sharma P C and Sharma H B. 2003. Rapid screening methodology for salt tolerance during germination and seedling emergence in Indian mustard (*Brassica juncea L.*). *Indian Journal of Plant Physiology* **8** (special issue): 363-67.

Soil Science Society of America, Committee on Terminology. 1960. Supplemental Report of Definitions Approved by the Committee on Terminology. *Soil Sci. Soc. Amer. Proceedings* **24**: 235-36.

Srivastava P, Bhattacharyya T and Pal D K. 2002.Significance of the formation of calcium carbonate minerals in the pedogenesis and management of cracking clay soils (Vertisols) of India. *Clays and Clay Minerals* **50**: 111-26.

Storer D A. 1984. A simple high sample volume ashing procedure for determination of soil organic matter. *Communications in Soil Science and Plant Analysis* **15** 759-72.

Suarez D L. 1981. Relationship between pH and SAR and an alternative method for estimating SAR of soil or drainage water. *Soil Sci. Soc. Am. J.* **45**: 469-75.

Suman B L. 1995. Usar bhumi sudhar: itihas ke aine se. Parti Bhumi Samachar. Jan.-March Issue. 27-31.

Swarup A. 1981. Effect of flooding on physico-chemical changes in sodic soils. *Z. PflanzenBodenk* **144**: 136-42.

Swarup A. 2000. Nutrient management for crop production in alkali lands. pp 99-108. *Salinity Management in Agriculture*. Gupta S K, Sharma S K and Tyagi N K (Eds.). Central Soil Salinity Research Institute, Karnal.

Swarup A and Yaduvanshi N P S. 2004. *Nutrient Stress Management for Sustaining Crop Production in Salt Affected Soils*. Bulletin No. **1/2004**. Pages 35. Central Soil Salinity Research Institute Karnal.

Szabolcs, I. 1979. Review of Research on Salt Affected Soils. Natural Resources Research **XV**. Pages 137. UNESCO, Paris.

Thokal R T, Gupta S K and Chauhan H S. 1992. Steady and transient modes of leaching saline soils. *J. Indian Soc. Soil Science* **40**: 219-34.

Tiwari A K, Pandey R N and Gupta S K. 1981. Transient behaviour of infiltration rate in alkali lands under reclamation. *J. Agric. Engineering* **18**: 21-8.

Tomar O S and Gupta Raj K. 1986. Performance of tree species in saline soils. *Ann. Report 1986*. pp 20-3. Central Soil Salinity Research Institute, Karnal.

Tomar O S, Gupta Raj K and Dagar J C. 1998. Afforestation techniques and evaluation of different tree species for waterlogged saline soils in semi-arid tropics. *Arid Soil Res. and Rehabilitation* **12**: 301-16.

Uppal H L. 1966. Reclamation of saline and alkali soils. pp 19.381-19-440. Proc. ICID Congress, Vol. II. New Delhi.

US Salinity Laboratory Staff. 1958. *Report of the Nomenclature Committee Appointed by the Board of Collaborators of the US Salinity Laboratory*. Pages 270. Soil Sci. Soc. of America and US Salinity Laboratory, Riverside. California.

Van Schilfgaarde J (Ed.). 1974. *Drainage for Agriculture*. Pages 700. Agronomy no. 17. American Society of Agronomy, Madison.

Verma S K and Gupta R K. 1989. Leaching behaviors of saline clay soil under two modes of water application. *Journal of the Indian Society of Soil Science* **37**: 903-08.

Walkley A J and Black I A. 1934. Estimation of soil organic carbon by chromic acid titration method. *Soil Science* **37**: 29-38.

Whitting L D and Janitzky P. 1963. Mechanism of formation of sodium carbonate in soil. *J. Soil Science* **14**: 322-33.

Wilson L G, Luthin J N and Biggar J W. 1961. *Drainage-salinity investigation of the Tulelake Lease Lands*. Bull. **776**. Pages 56. California Agric. Exp. Sta., CA, USA.

Yadav J S P. 1973. *Annual Report*. All India Coordinated Project on Water Management and Soil Salinity. Central Soil Salinity Research Institute, Karnal.

Yadav J S P and Girdhar I K. 1977. Salt translocation and crop growth under sprinkler and surface methods of irrigation in sodic soils. *Indian J. Agric. Science* **47**:397-400.

Yoder R E. 1936. A direct method of aggregate analysis of soils and a study of the physical nature of soil erosion. *J. Am. Soc. Agronomy* **28**: 337-51.

Zhawar V Kaur and Kaur K. 2018. Physiological and biochemical changes in plants under soil salinity stress: a review. pp 117-58. *Engineering Practices for Management of Soil Salinity: Agricultural, Physiological, and Adaptive Approaches*. Gupta S K, Goyal M R and Kmuar A. (Eds.). Apple Academic Press, USA.

Annexure - I

EQUIVALENT WEIGHT OF VARIOUS IONS, SALTS AND AMENDMENTS

Ions/Salts	Equivalent weight	Salts	Equivalent weight	Salts/amendments	Equivalent weight
Calcium	20.04	Magnesium chloride	47.62	Sodium carbonate	53.00
Magnesium	12.16	Magnesium sulphate	60.19	Sodium bicarbonate	84.01
Sodium	23.00	Magnesium chloride hexahydrate $MgCl_2\ 6H_2O$	101.7	Potassium chloride	74.56
Potassium	39.10	Magnesium sulphate heptahydrate $MgSO_4\ 7H_2O$	123.2	Potassium sulphate	87.13
Sulphate	48.03	Magnesium carbonate	42.16	Potassium carbonate	69.10
Chloride	35.46	Sodium chloride	58.45	Potassium bicarbonate	100.11
Carbonate	30.00	Sodium sulphate	71.03	Sulphur	16.03
Bicarbonate	61.01	Sodium carbonate	53.00	Sulphuric acid	49.04
Calcium chloride	55.50	Sodium bicarbonate	84.01	Aluminium sulphate	111.07
Calcium chloride dihydrate $CaCl_2\ 2H_2O$	73.5	Sodium chloride	58.45	Iron sulphate (ferrous)	139.01
Calcium sulphate	68.07	Sodium sulphate	71.03	Calcium carbonate	50.04
Calcium sulphate dihydrate $CaSO_4\ 2H_2O$	86.09				

Annexure - II

RECOMMENDED MAXIMUM CONCENTRATION (mg L^{-1}) OF TRACE ELEMENTS IN IRRIGATION WATERS[1]

Element	For waters used continuously on all soils	For use up to 20 years on fine textured soils of pH 6.0-8.5	Remarks
Aluminium (Al)	5.0	20.0	Can lower productivity in acid soils (pH 5.5), but more alkaline soils at pH 7.0 will precipitate the ion and eliminate any toxicity
Arsenic (As)	0.1	2.0	Toxicity to plants varies widely, ranging from 12 mg L^{-1}, for Sudan grass to less than 0.05 mg L^{-1} for rice
Beryllium (Be)	0.1	0.5	Toxicity to plants varies widely, ranging from 5 mg $L^{-1}$for kale to 0.5 mg $L^{-1}$for bush beans
Cadmium (Cd)	0.01	0.05	Toxic to beans, beets and turnips at concentrations as low as 0.1 mg L^{-1}in nutrient solutions. Conservative limits recommended due to its potential for accumulation in plants and soils to concentrations that may be harmful to humans.
Cobalt (Co)	0.05	5.0	Toxic to tomato plants at 0.1 mg L^{-1}in nutrient solution. Tends to be inactivated by neutral and alkaline soils
Chromium (Cr)	0.10	1.0	Not generally recognized as an essential growth element. Conservative limits recommended due to lack of knowledge of its toxicity to plants
Copper (Cu)	0.2	5.0	Toxic to a number of plants at 0.1 mg L^{-1} in nutrient solutions
Fluorine (F)	1.0	15.0	Inactivated by neutral and alkaline soils
Iron (Fe)	5.0	20.0	Not toxic to plants in aerated soils, but can contribute to soil acidification and loss of availability of essential phosphorus and

Element	For waters used continuously on all soils	For use up to 20 years on fine textured soils of pH 6.0-8.5	Remarks
			molybdenum. Overhead sprinkling may result in unsightly deposits on plants, equipment and buildings.
Lead (Pd)	5.0	10.0	Can inhibit plant cell growth at very high concentrations
Lithium (Li)	2.5	2.5	Tolerated by most crops up to 5 mg L^{-1}, mobile in soil. Toxic to citrus at low concentrations (0.075 mg L^{-1}). Acts similarly to boron.
Manganese (Mn)	0.2	10.0	Toxic to a number of crops at a few-tenths (Mn) to a few mg L^{-1}, but usually only in acid soils.
Molybdenum (Mo)	0.01	0.05	Not toxic to plants at normal concentration in soil and water. Can be toxic to livestock if forage is grown in soils with high concentrations of available molybdenum.
Nickel (Ni)	0.2	2.0	Toxic to a number of plants at 0.5 mg $L^{-1}$to 1.0 mg L^{-1}; reduced toxicity at neutral or alkaline pH.
Selenium (Se)	0.02	0.02	Toxic to plants at concentrations as low as 0.025 mg $L^{-1}$and toxic to livestock if forage is grown in soils with relatively high levels of added selenium. An essential element to animals but in very low concentrations.
Vanadium (V)	0.1	1.0	Toxic to many plants at relatively low concentrations
Zinc (Zn)	2.0	10.0	Toxic to many plants at widely varying concentrations; reduced toxicity at pH 6.0 and in fine textured or organic soils.

[1]The maximum concentration is based on a water application rate which is consistent with good irrigation practices (10,000 m^3 per hectare per year). If the water application rate greatly exceeds this, the maximum concentrations should be adjusted downward accordingly. No adjustment should be made for application rates less than 10,000 m^3 per hectare per year. The values given are for water used on a continuous basis at one site.

IMPORTANT CONVERSIONS

me	mg eq.wt^{-1}
ppm	mg L^{-1} or µg mL^{-1}
1 ppm	2.25 kg ha^{-1}
10,000 ppm	1 per cent
1 mg 100 g^{-1}	22.5 kg ha^{-1}
1 kg ha^{-1}	0.1 g m^{-2}
1 hectare	2.471 acres
1 acre	0.405 hectare
1 kg ha^{-1}	0.12 lbs acres^{-1}
Mesh number	16/ mm
Normality (Liquid)	(Specific gravity × Purity × 10)/ Equivalent weight
Normality (Solid)	Wt of salt per liter × Purity/ Equivalent weight
Grams per liter	Normality × Eq. Wt.
Organic matter	Organic carbon × 1.724
Optical density	2 – log T (T= % transmission)
Per cent by weight	Grams of solute/100 g of solution
1 Angstrom (Å)	10^{-8}cm or 10^{-10} m
10^{-8} cm or 10^{-10} m	1 nanometer or 1 milli-micrometer
°C / 5 (Temperature)	(°F-32) / 9
Protein (%)	Nitrogen (%)× 6.25
Velocity of speed	3 x 10^{-10} cm s^{-1}

CONVERSIONS FOR PRESSURE

psi	Atmosphere	mm H_2O	Kg cm^{-2}	mm Hg (Torr)	bar	Pa ($N\ m^{-2}$)	kPa	MPa
1	0.0681	703.8	0.0704	51.715	0.0689	6.895	6.895	0.0069
14.7	1	10,343	1.033	760	1.013	101.325	101.3	0.1013
0.0361	0.00246	25.4	0.00254	1.866	0.00249	248.8	0.249	0.00025
0.001421	0.000097	1	0.0001	0.0735	0.000098	9.8	0.0098	0.00001
0.01421	0.000967	10	0.001	0.735	0.00098	98	0.098	0.0001
0.0625	0.00425	43.986	0.0044	3.232	0.00431	431	0.431	0.00043
14.22	0.968	100,010	1	735.6	0.981	98,067	98.07	0.0981
0.4912	0.03342	345.7	0.0345	25.4	0.0339	3,386	3.386	0.00339
0.01934	0.001316	13.61	0.00136	1	0.001333	133.3	0.1333	0.000133
0.1934	0.01316	136.1	0.0136	10	0.01333	1,333	1.333	0.00133
0.0145	0.000987	10.21	0.00102	0.75	0.001	100	0.1	0.0001
14.504	0.987	10,210	1.02	750	1	100,000	100	0.1
0.000145	0.00001	0.102	0.00001	0.0075	0.00001	1	0.001	0.000001
0.14504	0.00987	102.07	0.0102	7.5	0.01	1,000	1	0.001
145.04	9.869	102,074	10.2	7500	10	1,000,000	1,000	1

ELEMENTAL CONVERSIONS

Percent or kg	Multiply by	Percent or kg
P_2O_5	0.437	P
K_2O	0.826	K
CaO	0.710	Ca
MgO	0.600	Mg
$CaCO_3$	0.400	Ca
NO_3-N	0.226	N
NH_4-N	0.777	N
SO_4	0.333	S
NH_3	0.893	N

■■■■